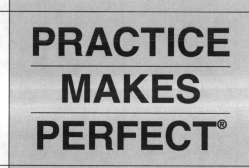

PRACTICE

MAKES

PERFECT®

Organic Chemistry

PRACTICE
MAKES
PERFECT®

Organic Chemistry

Marian DeWane

Thomas J. Greenbowe, PhD

Mc
Graw
Hill
Education

New York   Chicago   San Francisco   Athens   London   Madrid
Mexico City   Milan   New Delhi   Singapore   Sydney   Toronto

1 2 3 4 5 6 7 8 9   LHS   21 20 19 18 17

ISBN: 978-0-07-178986-8
MHID: 0-07-178986-3

e-ISBN: 978-0-07-178987-5
e-MHID: 0-07-178987-1

Contents

Introduction

Organic chemistry is a fascinating field, but it also can be very intimidating. The idea behind this book is to build a solid foundation in the basics. When taking an organic chemistry course, if you are not thoroughly familiar with these basics, you will be lost from the start. This book is designed to help ensure that you are ready.

The vast number and complexity of organic compounds can be daunting. Indeed, the majority of chemicals in the world are organic in nature. There is no way to cover every single compound, structure, or naming rule. However, organic chemicals can be thought about in a very orderly way to reduce the apparent complexity. Here, we have focused on the fundamentals you will need as you begin your study of organic chemistry and on making the "foreign language" of organic chemistry understandable by discerning its basic syntax. We have tried to adopt a step-by-step approach and to remove some of the mystery of organic chemistry. This book can serve either as an introduction to the subject or as a review of the foundational concepts that you need to be successful in an organic chemistry course for science majors and for pre-professional majors. Many of these concepts are discussed quickly in such a course, and unless you firmly grasp the basic concepts, you might struggle in the course. This foundation is also crucial if you are taking a survey of chemistry that includes a section on organic chemistry. You have taken the first step toward being on the right track.

Each chapter addresses a different type of organic compound and offers a detailed explanation. Problem-solving approaches, common mistakes, examples, and practice problems are all provided. The chapters build on each other and should be studied in sequence, but if you want to focus on a particular type of organic compound, each chapter can be studied independently.

Good luck with your studies!

Organic Chemistry

Carbon and the Study of Organic Chemistry

·1·

Organic chemistry involves studying most of the compounds made with the element carbon. Originally, only compounds made by living organisms were determined to be organic, but now many of these compounds can be synthesized in a laboratory. Essentially, the species not considered organic are the compounds carbon monoxide (CO) and carbon dioxide (CO_2), and compounds made with the ions carbonate (CO_3^{2-}) and cyanide (CN^-). There are millions of organic compounds, and this book will not attempt to teach about every single type of organic molecule. Rather, it will provide the foundational base for understanding the main groups of many common organic compounds.

The periodic table shows all the elements, their atomic numbers (number of protons present), and their average atomic mass.

Periodic Table of the Elements

Group	1	2	3	4	5	6	7	8	9	10	11	12	13	14	15	16	17	18
	1 / 1A																	18 / 8A
Period 1	1 **H** Hydrogen 1.008	2 / 2A											13 / 3A	14 / 4A	15 / 5A	16 / 6A	17 / 7A	2 **He** Helium 4.003
2	3 **Li** Lithium 6.941	4 **Be** Beryllium 9.012											5 **B** Boron 10.81	6 **C** Carbon 12.01	7 **N** Nitrogen 14.01	8 **O** Oxygen 16.00	9 **F** Fluorine 19.00	10 **Ne** Neon 20.18
3	11 **Na** Sodium 22.99	12 **Mg** Magnesium 24.31	3 / 3B	4 / 4B	5 / 5B	6 / 6B	7 / 7B	8	9 / 8B	10	11 / 1B	12 / 2B	13 **Al** Aluminum 26.98	14 **Si** Silicon 28.09	15 **P** Phosphorus 30.97	16 **S** Sulfur 32.07	17 **Cl** Chlorine 35.45	18 **Ar** Argon 39.95
4	19 **K** Potassium 39.10	20 **Ca** Calcium 40.08	21 **Sc** Scandium 44.96	22 **Ti** Titanium 47.88	23 **V** Vandium 50.94	24 **Cr** Chromium 52.00	25 **Mn** Manganese 54.94	26 **Fe** Iron 55.85	27 **Co** Cobalt 58.93	28 **Ni** Nickel 58.69	29 **Cu** Copper 63.55	30 **Zn** Zinc 65.39	31 **Ga** Gallium 69.72	32 **Ge** Germanium 72.59	33 **As** Arsenic 74.92	34 **Se** Selenium 78.96	35 **Br** Bromine 79.90	36 **Kr** Krypton 83.90
5	37 **Rb** Rubidium 85.47	38 **Sr** Strontium 87.62	39 **Y** Yttrium 88.91	40 **Zr** Zircanium 91.22	41 **Nb** Niobium 92.91	42 **Mo** Molybdenum 95.94	43 **Tc** Technetium (98)	44 **Ru** Ruthenium 101.1	45 **Rh** Rhodium 102.9	46 **Pd** Palladium 106.4	47 **Ag** Silver 107.9	48 **Cd** Cadmium 112.4	49 **In** Indium 114.8	50 **Sn** Tin 118.7	51 **Sb** Antimony 121.8	52 **Te** Tellurium 127.6	53 **I** Iodine 126.9	54 **Xe** Xenon 131.3
6	55 **Cs** Cesium 132.9	56 **Ba** Barium 137.3	57 **La** Lanthanum 138.9	72 **Hf** Hafnium 178.5	73 **Ta** Tantalum 180.9	74 **W** Tungsten 183.9	75 **Re** Rhenium 186.2	76 **Os** Osmium 190.2	77 **Ir** Iridium 192.2	78 **Pt** Platinum 195.1	79 **Au** Gold 197.0	80 **Hg** Mercury 200.5	81 **Ti** Thallium 204.4	82 **Pb** Lead 207.2	83 **Bi** Bismuth 209.0	84 **Po** Polonium (210)	85 **At** Astatine (210)	86 **Rn** Radon (222)
7	87 **Fr** Francium (223)	88 **Ra** Radium (226)	89 **Ac** Actinium (227)	104 **Rf** Rutherfordium (257)	105 **Db** Dubnium (260)	106 **Sg** Seaborgium (263)	107 **Bh** Bohrium (262)	108 **Hs** Hassium (265)	109 **Mt** Meitnerium (266)	110 **Ds** Darmstadtium (269)	111 **Rg** Roentgenium (272)	112 **Cn** Coperisium (285.17)	113 **Nh** Nihonium (284.18)	114 **Fl** Flerovium (289.19)	115 **Mc** Moscovium (288.19)	116 **Lv** Livermorium (293)	(117) **Ts** Tennessine (294)	118 **Og** Oganesson (294)

10 ← Atomic number
Ne
Neon
20.18 ← Atomic mass

Metals

Metalloids

Nonmetals

58 **Ce** Cerium 140.1	59 **Pr** Praseodymium 140.9	60 **Nd** Neodymium 144.2	61 **Pm** Promethium (147)	62 **Sm** Samarium 150.4	63 **Eu** Europium 152.0	64 **Gd** Gadolinium 157.3	65 **Tb** Terbium 158.9	66 **Dy** Dysprosium 162.5	67 **Ho** Holmium 164.9	68 **Er** Erbium 167.3	69 **Tm** Thulium 168.9	70 **Yb** Ytterbium 173.0	71 **Lu** Lutetium 175.0
90 **Th** Thorium 232.0	91 **Pa** Protactinium (231)	92 **U** Uranium 238.0	93 **Np** Neptunium (237)	94 **Pu** Plutonium (242)	95 **Am** Americium (243)	96 **Cm** Curium (247)	97 **Bk** Berkelium (247)	98 **Cf** Californium (249)	99 **Es** Einsteinium (254)	100 **Fm** Fermium (253)	101 **Md** Mendelevium (256)	102 **No** Nobelium (254)	103 **Lr** Lawrencium (257)

This looks like there can be a huge number of organic compounds combining all these elements with carbon. (There actually are a huge number of organic compounds, but for a different reason.) In fact, most organic compounds use the following elements on the periodic table.

Group — Principal elements involved in organic chemistry

So why can there be so many organic compounds? Carbon has the unique property of being able to form chains—sometimes huge chains—increasing the numbers of possible compounds. Carbohydrates, including sugars, fats, and some polymers, are examples of molecules with long chains of carbons.

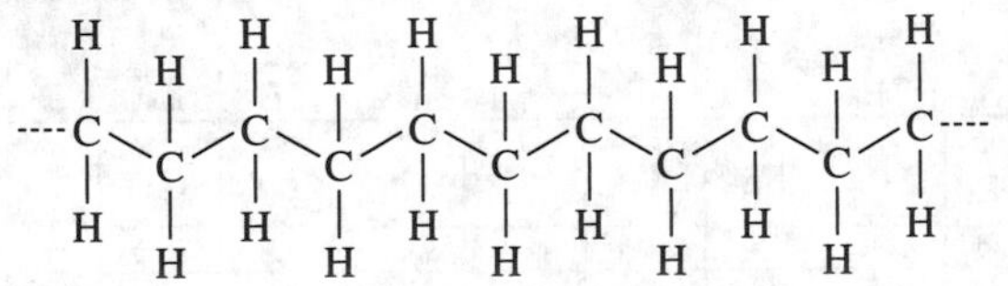

Carbon chains can also hook one end to the other end and form rings. The sugar glucose has the formula $C_6H_{12}O_6$.

Glucose

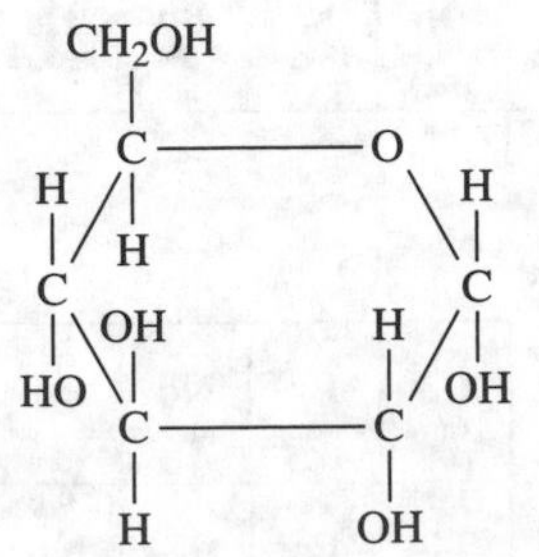

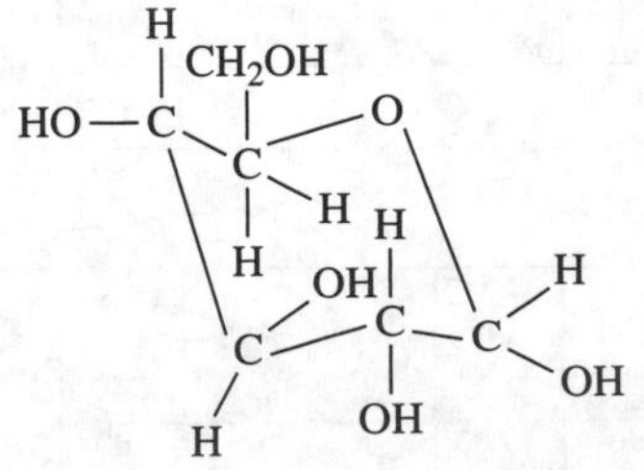

Glucose

A few concepts learned in an inorganic chemistry course are needed in the study of organic chemistry. Specifically, it is necessary to be familiar with atomic structure, molar mass, isomers, Lewis structures, hybridization, the forces holding one molecule to another molecule, and molecular polarity. The following sections provide a basic review of these concepts.

Atomic Structure

The three main particles in the atom are protons, neutrons, and electrons. Each is important for different atomic properties. The neutrons and protons are both found in the nucleus of the atom and are collectively called nucleons. The nucleus of an atom can be represented by a small sphere within a much larger sphere. Electrons are found outside the nucleus but within the larger sphere.

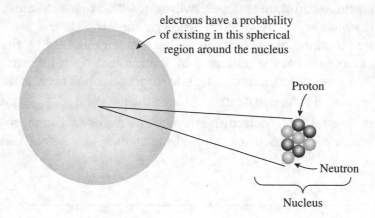

electrons have a probability of existing in this spherical region around the nucleus

Proton

Neutron

Nucleus

Protons and neutrons each have a mass of approximately 1 atomic mass unit (amu), although the neutron has the slightly larger mass. An atomic mass unit is equal to 1.661×10^{-27} kilograms, the recognized International System unit (SI) of mass. Kilograms can be converted to grams using the conversion factor of $\dfrac{1000 \text{ grams}}{1 \text{ kilogram}}$, so 1.000 amu is 1.661×10^{-24} grams.

$$1.000 \text{ amu} \times \frac{1.661 \times 10^{-27} \text{ kg}}{1.0 \text{ amu}} \times \frac{1000 \text{ g}}{1.0 \text{ kg}} = 1.661 \times 10^{-24} \text{ g}$$

In comparison, the electron has a mass of 0.00054 amu. The electron mass is so small that it does not significantly affect the overall mass of the atom. Using the rules of significant figures and rounding addition to the least measured place value, the result of the addition is rounded to the nearest tenth; 1.0 amu + 0.00054 amu = 1.0 amu. Even for one of the larger elements with many electrons, the electron mass does not change the total mass by much. Masses listed in the periodic table do include the electron masses, and different versions of the table will have different numbers of significant figures. In the case of hydrogen, the mass of the electron is significant as only a single proton is present. Since most of the mass of an atom is due to the masses of protons and neutrons located in the nucleus, to determine the mass of an atom, then, we must know the number of protons and neutrons.

All atoms of the same element have the same number of protons. This number is called the atomic number and is represented by the symbol Z. For instance, carbon is element atomic number 6. It has six protons. Hydrogen is element 1, and it has one proton. Since the mass number (A) is the sum of the number of protons and neutrons, we can write the formula $p + n =$ mass number, where p is the proton number and n is the neutron number, then rearranged using Z for protons and N for the total neutron number $Z + N = A$. If the proton number and a mass number are known, this can be rearranged to the form $A - Z = N$ to solve for the number of neutrons. Carbon atoms with a mass of 12 amu have six neutrons, whereas carbon atoms with a mass of 13 amu have seven neutrons, and carbon atoms with a mass of 14 amu have eight neutrons. When atoms of the same element have different masses, because of differing numbers of neutrons, they are called isotopes of each other. Carbon has fifteen known isotopes, although the three common isotopes are those with a masses of 12 amu, 13 amu, and 14 amu, respectively. We can write nuclide symbols for the isotopes of carbon: ^{12}C, ^{13}C, and ^{14}C. In this notation, the mass number is written as a superscript to the left of the chemical symbol of the element. (In the equations below p represents protons and Z represents the total number of protons; n represents neutrons and N represents the total number of neutrons.)

$$p + n = \text{mass number, or } Z + N = A$$

$$\text{mass number} - p = n, \text{ or } A - Z = N$$

Hydrogen has three common isotopes: hydrogen with a mass of 1 amu, a mass of 2 amu, and a mass of 3 amu, which are represented by the symbols ^1H, ^2H, and ^3H, respectively. It is also acceptable to write the name of the element with a hyphen followed by the mass. For hydrogen, then, the different isotopes can be indicated by writing hydrogen-1, hydrogen-2, or hydrogen-3. Since hydrogen-1 has one proton, subtracting one from each mass reveals that ^1H has no neutrons, ^2H has one neutron, and ^3H has two neutrons. Using the notation $^{\text{mass number}}_{\text{proton number}}$ atomic symbol, or $^A_Z X$, sets up a subtraction equation to determine the neutron number. The setup ^1_1H indicates $1 - 1 = 0$ neutrons, ^2_1H indicates $2 - 1 = 1$ neutron, and ^3_1H indicates $3 - 1 = 2$ neutrons.

EXERCISE

1·1

Determine the number of protons in each of the following elements. (Hint: use the periodic table above.)

1. silicon   _____

2. oxygen   _____

3. phosphorus   _____

4. chlorine   _____

5. nitrogen   _____

EXERCISE

1·2

Determine the number of neutrons in each of the following atoms.

1. silicon-29   _____

2. silver-109   _____

3. oxygen-16   _____

4. oxygen-17   _____

5. potassium-41   _____

EXERCISE

1·3

Write the atomic symbol $_Z^A X$ for atoms with the following numbers of protons and neutrons.

1. 19 protons and 20 neutrons   _____

2. 14 protons and 14 neutrons   _____

3. 14 protons and 16 neutrons   _____

4. 17 protons and 18 neutrons   _____

5. 5 protons and 6 neutrons   _____

EXERCISE

1·4

How many protons and how many neutrons do these atoms have?

1. ^{24}Al   _____

2. ^{60}Co   _____

3. ^{19}F   _____

4. ^{13}C   _____

5. ^{15}N   _____

On the periodic table, the mass of hydrogen is 1.008 amu. This is because the mass of an element on the periodic table is a weighted average of all the natural isotopes of that element. To determine a weighted average mass, the percentage of occurrence of each isotope in a sample is needed, (usually that found in nature) as well as the actual masses, not the rounded whole-number masses. Hydrogen has natural isotopes, ^{1}H (99.9885%), ^{2}H (0.01155%), and ^{3}H (4×10^{-15}%, extremely small).

The actual mass of hydrogen-1 is 1.00782 amu, that of hydrogen-2 is 2.01410 amu, and that of hydrogen-3 is 3.0160 amu. When masses of this precision are used in average mass calculations, the electron masses have been included.

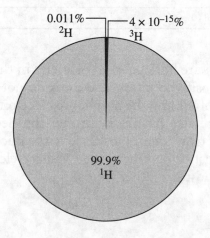

To determine the weighted average mass, the following general formula is used: [(percentage/100) × mass] + [(percentage/100) × mass] + [(percentage/100) × mass] + · · · = weighted average mass. For hydrogen this would be [(99.9885/100) × 1.00782 amu] + [(0.0115/100) × 2.01410 amu] + [(4 × 10^{-15}/100) × 3.0160 amu] = 1.0079, or 1.008 amu, the mass of hydrogen displayed on the periodic table. Note that there is no individual hydrogen atom with this exact mass.

EXERCISE
1·5

Determine the average atomic mass for each of the following elements.

1. Magnesium, with three naturally occurring isotopes: ^{24}Mg with an actual mass of 23.9850 amu and an abundance of 78.99%, ^{25}Mg with an actual isotopic mass of 24.9858 and an abundance of 10.00%, and ^{26}Mg with an actual isotopic mass of 25.9826 and an abundance of 11.01%.

_____

_____

2. Silver, with two naturally occurring isotopes with masses of 107 (106.9051 amu) and 109 (108.9048 amu) and respective abundances of 51.84% and 48.16%.

_____

_____

3. Silicon, with three isotopes with masses 28 (27.9769 amu), 29 (28.9765 amu), and 30 (29.9737) and respective abundances of 92.23%, 4.67%, and 3.10%.

_____

_____

4. Boron, with two naturally occurring isotopes with masses of 10 (10.013 amu) and 11 (11.009 amu) and respective abundances of 20.00% and 80.00%.

_____

_____

5. Neon, with three naturally occurring isotopes of 20 (19.992 amu), 21 (20.994 amu), and 22 (21.991 amu) and respective abundances of 90.48%, 0.27%, and 9.25%.

_____

_____

The electron and proton are both charged particles in the atom. The electron is negatively charged and the proton is positively charged. Atoms do not normally carry a charge and are neutral, so the number of protons and the number of electrons must be equal. The atom carbon has six protons and therefore also has six electrons. The electrons are outside the nucleus in locations called orbitals. Electrons in orbitals close to the nucleus are called core electrons, and the electrons in the outermost orbitals are called valence electrons. The valence electrons are the electrons involved in bonding to other atoms.

EXERCISE
1·6

How many total electrons are there in each of the following atoms?

1. sulfur _____

2. oxygen _____

3. nitrogen _____

4. chlorine _____

5. hydrogen _____

When the electrons and protons are not equal in number, the atom is called an ion. If there are more protons than electrons, the overall charge is positive and the atom is called a cation. If there are more electrons than protons, the overall charge is negative and the atom is called a anion. The proton number minus the electron number will indicate the charge. Oxygen with eight protons and 10 electrons is the anion O^{2-}. The charge is written after the symbol as a superscript. The number is written first and the charge (positive or negative) follows. If the number is 1, it is omitted and only the charge (+ or −) is written.

EXERCISE
1·7

Determine the number of protons and the number of electrons in each of the following ions.

1. C^{4-} _____

2. N^{3-} _____

3. Na^+ _____

4. Ca^{2+} _____

5. S^{2-} _____

EXERCISE
1·8

Determine if each of the following species is an ion or atom and write the appropriate symbol.

1. 15 protons and 18 electrons _____

2. 13 protons and 10 electrons _____

3. 30 protons and 28 electrons _____

4. 10 protons and 10 electrons _____

5. 9 protons and 10 electrons _____

Knowing the number of outer electrons for each atom is important in drawing Lewis structures, an important skill in organic chemistry. Lewis structures are used to help determine the types of bonding between atoms in a molecule. The total electrons minus the core electrons will equal the valence electrons. For elements in the first row (rows are called periods), the number of core electrons is zero; for elements in the second period, the number of core electrons is two, and for elements in the third period, the number of core electrons is ten. Because the proton number and electron number are equal in neutral atoms, the number of core electrons is the number of the last element of the period before. For hydrogen, element 1 in the first period, $1 - 0 = 1$ valence electron. For potassium, element 19, the last element number of the row before is 18, so the number of valence electrons is $19 - 18 = 1$. This trend indicates that for the atoms in the first column of the periodic table, the number of valence electrons is one.

EXERCISE

1·9

Determine the number of valence electrons for each of the following atoms.

1. carbon   _____

2. nitrogen   _____

3. oxygen   _____

4. chlorine   _____

5. sulfur   _____

Molar Mass

The mass of 1 molecule of a compound is *very* small. For instance, methane is composed of one carbon atom and four hydrogen atoms. Looking at the periodic table, the mass of a carbon atom is 12.01 amu and the mass of each hydrogen atom is 1.008 amu, so the total for methane is $(1 \times 12.01) + (4 \times 1.008) = 16.04$ amu. Given the conversion factor ratio 1 amu $= 1.661 \times 10^{-24}$ g, the mass of one molecule is 16.04 amu $\times \dfrac{1.661 \times 10^{-24} \text{ g}}{1 \text{ amu}} = 2.66 \times 10^{-23}$ g. This mass is too small to be measured on a balance in a chemistry laboratory. Instead, a number of molecules are massed together. The unit used for a large number of atoms or molecules or similar species is called a mole.

The term "mole," much like the term "dozen," represents a specific number of items. In chemistry, a mole consists of 6.022×10^{23} objects such as atoms, molecules, or other particles. This number, which is called Avogadro's number, is based on the number of atoms of carbon in 12.00 grams of the carbon-12 isotope. The mass of 6.022×10^{23} molecules is called molar mass. When the mass of one molecule of methane is multiplied by Avogadro's number, the result is the mass of one mole, a mass measurable using a laboratory scale.

$$(2.66 \times 10^{-23} \text{ g}) \times (6.022 \times 10^{23}) = 16.04 \text{ g/mol}$$

The mass in atomic mass units of one molecule is the same number as the mass in grams of a mole! Hence, the steps in adding and using molar mass are the same as those in adding and using the mass of one molecule. The difference is only in the unit.

The steps in calculating molar mass are first to determine how many atoms of each element are in the formula, next to multiply the mass of each element by the number of atoms of that element present in the formula, and then to add those masses together. For example, the formula $C_{12}H_{22}O_{11}$ of sucrose would be broken into 12 carbon, 22 hydrogen, and 11 oxygen atoms. Using the masses from the periodic table, and using them as molar masses in grams, this translates into the equation

$$(12 \times 12.01 \text{ g}) + (22 \times 1.008 \text{ g}) + (11 \times 16.00 \text{ g}) = 342.3 \text{ g/mol}$$

A basic format for such an equation is

$$(\text{number of A} \times \text{mass of A}) + (\text{number of B} \times \text{mass of B}) + (\cdots) = \text{molar mass}$$

EXERCISE

1·10

Find the molar mass for each of the following compounds.

1. C_4H_{10} _____

2. C_4H_9OH _____

3. CH_3NH_2 _____

4. CH_4O _____

5. C_2H_6O _____

6. C_3H_8O _____

7. $C_3H_6Cl_2$ _____

8. C_2H_2 _____

9. $C_6H_{12}O_6$ _____

10. C_2H_5OH _____

Lewis Structures

Chemical formulas do not indicate the way atoms are joined or arranged in space. A Lewis structure helps to show the way the atoms are arranged and bonded together. Since carbon is the base of all organic chemistry, it is critical to know that while carbon has six electrons, it has four outer (valence) electrons that can be involved in bonding.

When bonds form in the covalent bonds of organic molecules, electrons are shared between the atoms in the bond. The octet rule states when atoms form a compound, they form bonds in such a way that each atom (primarily) has eight electrons in its valence shell. Lewis also introduced the idea that each bond consisted of a pair of electrons, with usually one electron being contributed by each atom forming the bond. With four valence electrons, carbon needs four more electrons to have eight electrons (that is, an octet of electrons) in its outer shell. In the case of carbon, if all four of the electrons form a bond, this means that four bonds form. Each bond represents two electrons, one from carbon and one from the atom it is sharing with. Hydrogen has only one valence electron and does not follow the octet rule. It can form only one bond and therefore can have a maximum of only two electrons. For the principal atoms found in organic compounds, the number of valence electrons and the structure with valence electrons are as follows:

Element	Symbol	Number of valence electrons	Structure with valence electrons
Hydrogen	H	1	H·
Carbon	C	4	·Ċ·
Nitrogen	N	5	·N̈·
Oxygen	O	6	:Ö·
Chlorine	Cl	7	:C̈l·
Sulfur	S	6	:S̈·
Bromine	Br	7	:B̈r·

The arrangement of the eight electrons with four bonds around the carbon could be four single bonds. However, sometimes atoms, especially of carbon, can share more than one pair of electrons. The following diagrams show each bond with two shared electrons; a single bond with six shared electrons, a double bond with four shared electrons, or two double bonds, each with four electrons.

$$-\overset{\mid}{\underset{\mid}{C}}- \qquad -C\equiv \qquad {>}C{=}$$

In organic chemistry, the type and number of bonds are often indicated in the name of the compound. This will be covered in the chapters on the different compounds. Determining the structure from the formula is often more difficult. If the compound has only carbon and hydrogen atoms, the ratio of one to the other will help indicate the type of bonds present. This will also be covered in each chapter where appropriate. In general, for organic compounds (ions will not be addressed here) with hydrogen and carbon present, the steps to creating a molecular structure are:

◆ Draw the Lewis structure of each atom involved in the compound, and determine the total number of valence electrons to be in the structure.
◆ Arrange the atoms in a preliminary structure with the carbons arranged in a chain. Place two electrons in each bond.
◆ For common elements involved with carbon to make simple organic molecules, count how many electrons are needed to follow the octet rule. (As we have noted, hydrogen is an exception to the octet rule.) Does the number in the diagram equal the total number of valence electrons that you determined in the first step? If so, only single bonds are present. If not, compare the number you used in the original diagram with the number of electrons present. For every difference of two electrons, a multiple bond is present. Each pair of shared electrons can be represented with a bond line.

For example, C_3H_8 has 3 carbon atoms and 6 hydrogen atoms. Each carbon has 4 valence electrons and each hydrogen has one electron, for a total of 14 valence electrons.

$$\cdot\dot{C}\cdot \quad \cdot\dot{C}\cdot \quad \cdot\dot{C}\cdot \quad H\cdot \quad H\cdot \quad H\cdot \quad H\cdot \quad H\cdot \quad H\cdot$$

Arrange the carbons in a chain with two electrons between the adjacent carbon atoms. Arrange the hydrogen atoms around the carbons so that there are four bonds on each carbon.

$$
\begin{array}{ccc}
H & H & H \\
H\cdot\cdot\dot{C}\cdot\cdot\dot{C}\cdot\cdot\dot{C}\cdot\cdot H \\
H & H & H
\end{array}
$$

All atoms except hydrogen have eight electrons around and each hydrogen has two. Counting up the number of electrons present, this is equal to the number available. Substituting a bond line for each pair of shared electrons results in the final structure.

$$
\begin{array}{ccccc}
 & H & & H & & H \\
 & | & & | & & | \\
H & - & C & - & C & - & C & - & H \\
 & | & & | & & | \\
 & H & & H & & H
\end{array}
$$

Another example is C_2H_4. The two carbon atoms and the 4 hydrogen atoms have a total of 12 valence electrons.

$$\cdot\dot{C}\cdot \quad \cdot\dot{C}\cdot \quad H\cdot \quad H\cdot \quad H\cdot \quad H\cdot$$

Arrange the two carbons next to each other with two electrons between them. To complete this diagram with all single bonds would take six hydrogen atoms, but only four are present.

$$
\begin{array}{cc}
H & H \\
H\cdot\cdot\dot{C}\!:\!\dot{C}\!\cdot & \longleftarrow \text{ need 2 more } H\cdot \\
H
\end{array}
$$

The difference of two indicates that a double bond must be present between the carbons. Place four electrons between the carbons, and then place two hydrogen atoms and their electrons in the remaining positions. This adds up to twelve electrons. Substitute a bond line for each pair of shared electrons.

$$
\begin{array}{ccc}
H & & H \\
\ddot{C}\!:\!:\!\ddot{C} & \quad & H\diagdown C\!=\!C\diagup H \\
H & & H \diagup \diagdown H
\end{array}
$$

EXERCISE

1·11

Draw the Lewis structure for each of the following compounds.

1. CH_4

2. C_3H_8

3. C_2H_2

4. C_3H_4

5. C_2H_6

Hybridization Around the Carbon Atom

In the structures drawn in organic chemistry, the carbon atom will use either four, three, or two domains around the atom for bonding. Models of atoms suggest electrons are directed at 90 degrees in an atom. However, it is known that electron pair bonds tend to distribute themselves evenly in space around the "central" atom. To solve this dilemma, scientists introduced the term hybridization which imagines blending of the atomic orbitals in the process of forming bonds. If four carbon electrons are used independently in bonding, the carbon is said to have sp^3 hybridization; if only three are used, the carbon has sp^2 hybridization; and if two are used, the carbon has sp hybridization. The hybridization determines the shape around the carbon: a tetrahedral shape for sp^3 hybridization, a trigonal planar shape for sp^2 hybridization, and a linear structure for sp hybridization.

sp^3 4 domains sp^2 3 domains sp 2 domains

Structural Isomers

A chemical formula indicates only the number of each type of atom present. It does not indicate the structure. Structural isomers are compounds with the same number of each atom but different structures and different compound names. This is an important concept in organic chemistry, given the many possible ways in which carbon atoms can link together. An example is C_4H_{10}. This formula can be butane or 2-methyl propane, which is also called isobutane. The prefix iso- stands for "isomer of." The specifics of naming will be covered in each chapter where naming is relevant.

Butane   Isobutane

  Both structures have the same number of each atom but different arrangement of the atoms. Butane and isobutane are different compounds with different properties. Using a different type of formula in organic chemistry is helpful for indicating the structure so that the actual molecule can be known. In the case of butane, there are two isomers. As the number of carbons increases, the number of isomers increases as well. For example, in the case of pentane, adding just one additional carbon in the chain gives rise to three isomers.

Forces Between Molecules

The forces between molecules determine physical characteristics such as melting points and boiling points. Organic compounds have covalent bonds in the molecule (pair bonds in the Lewis structures). These bonds are much stronger than the forces holding one molecule to another.

Intermolecular forces both hold one molecule to another to form a solid and cause the molecules of liquids to stay close to each other. The amount of energy needed to overcome the force holding the molecules together in a liquid and enable them to separate and escape as gas molecules is an indication of the strength of the force. A force found in all organic molecules is the dispersion force. The strength of the dispersion force is mainly determined by the total number of electrons distributed in the molecule, not just by the valence electrons. Since electrons are in constant motion, at any one time the distribution of electrons can temporarily be uneven within the molecule, creating a temporary dipole, or area of uneven charge. Such a molecule is said to be polarized. The more electrons present, the larger the electron cloud, and the easier it is to polarize the molecule. The resulting temporary dipoles create an attraction to another molecule. If a compound contains only carbon and hydrogen atoms, then only dispersion forces are present as C-H bonds are relatively non-polar.

Another force that is often stronger than a single dispersion force is a dipole force. Dipole-dipole forces occur when permanent dipoles exist in the molecule. When there is an unsymmetrical distribution of electron-poor regions and electron-rich regions, permanent dipoles occur. To determine whether a permanent dipole exists, there are two things to consider. The first is whether the bonds inside the molecule are polar or nonpolar. To determine whether the bonds are polar, it is necessary to consider differences in electronegativity between the atoms. In the following Periodic Table, electronegativity values for each element are listed.

Increasing electronegativity →

Increasing electronegativity ↑

1A	2A	3B	4B	5B	6B	7B	8B	8B	8B	1B	2B	3A	4A	5A	6A	7A	8A
H 2.1																	
Li 1.0	Be 1.5											B 2.0	C 2.5	N 3.0	O 3.5	F 4.0	
Na 0.9	Mg 1.2											Al 1.5	Si 1.8	P 2.1	S 2.5	Cl 3.0	
K 0.8	Ca 1.0	Sc 1.3	Ti 1.5	V 1.6	Cr 1.6	Mn 1.5	Fe 1.8	Co 1.9	Ni 1.9	Cu 1.9	Zn 1.6	Ga 1.6	Ge 1.8	As 2.0	Se 2.4	Br 2.8	Kr 3.0
Rb 0.8	Sr 1.0	Y 1.2	Zr 1.4	Nb 1.6	Mo 1.8	Tc 1.9	Ru 2.2	Rh 2.2	Pd 2.2	Ag 1.9	Cd 1.7	In 1.7	Sn 1.8	Sb 1.9	Te 2.1	I 2.5	Xe 2.6
Cs 0.7	Ba 0.9	La-Lu 1.0-1.2	Hf 1.3	Ta 1.5	W 1.7	Re 1.9	Os 2.2	Ir 2.2	Pt 2.2	Au 2.4	Hg 1.9	Ti 1.8	Pb 1.9	Bi 1.9	Po 2.0	At 2.2	
Fr 0.7	Ra 0.9																

If the same two atoms are bonded to each other, such as a carbon to a carbon, the result is zero (2.5 – 2.5 = 0). When a hydrogen atom bonds to a carbon atom, there is a difference in electronegativity, but it is low enough for us to consider the C-H bond nonpolar (2.5 – 2.1 = 0.4). Generally, differences in electronegativity less than 0.4 characterize nonpolar bonds. In contrast, when an oxygen atom bonds to a hydrogen atom, there is an electronegativity difference of 1.4 (3.5 – 2.1 = 1.4). This O-H bond is a polar covalent bond. A difference between 0.5 and 2.0 is considered to indicate polarity. (A difference in electronegativity greater than 2.0 reflects an ionic bond but is not a factor in organic molecules.)

0.0 – 0.4   nonpolar bonds
0.5 – 2.0   polar bonds

Determine whether each of the following bonds is polar or nonpolar.

1. N – O _____

2. Si – O _____

3. C – S _____

4. O – S _____

5. P – F _____

However, this is not all that must be considered. The second thing to consider in determining whether a permanent dipole exists is whether a molecule has symmetry. A molecule can have polar bonds and be a nonpolar molecule. How? All the dipoles cancel each other, in a truly symmetrical molecule. Take carbon dioxide. Each carbon-oxygen bond has a polarity of $3.5 - 2.5 = 1.0$. When the Lewis structure is drawn as a 3D structure, carbon dioxide is a linear molecule. Since oxygen is more electronegative compared to carbon, each oxygen atom has a partial negative charge and the carbon atom has a partial positive charge.

$$\ddot{O} = C = \ddot{O} \qquad\qquad \overset{\delta^-}{O} = \overset{\delta^+}{C} = \overset{\delta^-}{O}$$
$$\uparrow \quad \uparrow \qquad\qquad \leftarrow \quad \longrightarrow$$
$$1.0 \quad 1.0$$

each side has a 1.0 polarity, but the molecule has a symmetrical distribution, and the dipoles cancel, rendering the molecule nonpolar.

In addition to the polarity of bonds, the shape of the molecule must be considered in determining whether a molecule is polar.

$$H - \ddot{N} - H \qquad\qquad H^{\cdots} \overset{\ddot{N}}{\underset{H}{\diagup}} {\diagdown} H$$
$$\underset{H}{|}$$

In ammonia, the nitrogen atom has an electronegativity value of 3.0 and the hydrogen has a value of 2.1. The difference in electronegativity between these two atoms is 0.9, which identifies the N-H bond as polar covalent. Because nitrogen has a higher electronegativity compared to hydrogen, nitrogen will pull the electron density in the N-H bond closer. This creates a partial negative charge on the nitrogen atom and partial positive charges on the hydrogen atoms. Overall, ammonia is a polar molecule. Ammonia has a a net dipole moment. The partial negative end of the dipole moment in one ammonia molecule is attracted to the partial positive end of the dipole moment of another ammonia molecule.

Hydrogen bond

This polarity also determines whether a liquid substance is miscible in water. In general, polar molecules are miscible in water; in other words, they mix together and do not separate into two solutions. If the molecule is nonpolar and has only dispersion forces as its intermolecular force, it

Corn oil and water do not mix. The corn oil is the top layer and water is the bottom layer.

Baby oil (top layer) and water (bottom layer) do not mix.

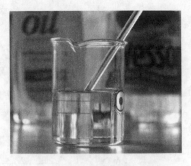

Corn oil and baby oil mix to form a solution.

will not be miscible in water. For example, corn oil and water do not mix. A nonpolar liquid will be miscible in another nonpolar solvent. For example, corn oil and water do not mix. A nonpolar molecule will be miscible in another nonpolar solvent. For example, baby oil mixes with corn oil.

Some molecules can form super dipole-dipole forces called hydrogen bonds. It can sound confusing, but these are not bonds inside a molecule; rather, they are forces between one molecule and another. For a hydrogen bond to occur, a molecule must have a "FON," which means that the molecule must have a fluorine, an oxygen, or a nitrogen atom attached to a hydrogen atom. F, O, and N have high electronegativity values and are relatively small atoms. The F, O, or N in one molecule is attracted to the hydrogen in another molecule. For example, methanol (CH$_3$OH) has an electron-rich, small, electronegative oxygen that is strongly attracted to the electropositive hydrogen in another molecule of methanol. The hydrogen bond is indicated by a dashed line. Be aware, however, that even though an F, O, or N is necessary for hydrogen bonding to occur, hydrogen bonding does not occur every time an F, O, or N and hydrogen are present.

EXERCISE
1·13

For each of the following molecular structures, determine the electronegativity difference for each bond, and then determine whether the molecule is polar. Explain how you determined whether the molecule is polar or not.

1. H—Ö:
      |
      H

_____

_____

2. H—N̈—H
       |
       H

_____

_____

3.

$$\underset{\underset{\displaystyle H}{|}}{\overset{\overset{\displaystyle H}{|}}{H-C}} - \underset{\underset{\displaystyle H}{|}}{\overset{\overset{\displaystyle H}{|}}{C}} - H$$

_____

_____

4.

$$\underset{\underset{\displaystyle H}{|}}{\overset{\overset{\displaystyle H}{|}}{H-C}} - C \underset{\overset{\displaystyle \ddot{O}}{\diagdown}}{\diagup \ddot{O}} \diagdown H$$

_____

_____

5.

$$\underset{\underset{\displaystyle H}{|}}{\overset{\overset{\displaystyle H}{|}}{H-C}} - \overset{\overset{\displaystyle :\ddot{O}:}{\|}}{C} - \underset{\underset{\displaystyle H}{|}}{\overset{\overset{\displaystyle H}{|}}{C}} - H$$

_____

_____

EXERCISE
1·14

Put an x in each box to indicate the types of forces present in the following molecules.
These formulas represent the same molecules whose structures appeared in Exercise 1-13.

MOLECULE	DISPERSION FORCE	DIPOLE-DIPOLE FORCE	HYDROGEN BOND
1. H_2O	_____	_____	_____
2. NH_3	_____	_____	_____
3. C_2H_6	_____	_____	_____
4. $C_2H_4O_2$	_____	_____	_____
5. C_3H_6O	_____	_____	_____

Which was more helpful in determining the forces present, the formula or the structure? Explain.

_____

_____

_____

Alkanes

·2·

Organic chemistry starts with a study of the simplest hydrocarbons, the alkanes. Alkanes are organic compounds having the general formula $C_nH_{(2n+2)}$, where n is the number of carbons. For two carbons, there are $2(2) + 2 = 6$ hydrogen atoms. Alkanes by definition are hydrocarbons that are made only of carbon atoms saturated with hydrogen atoms. Being saturated means that all the bonds between carbon atoms are single bonds sharing two electrons. Since all bonds are C-H or C-C, and only dispersion forces exist between molecules, all alkanes are nonpolar molecules and are insoluble in water. When other atoms are present, the substance can no longer be called a simple alkane.

The alkanes with lower molar masses are gases at room temperature and 1 atmosphere of pressure. Many of these are used as fuels. Familiar fuels include methane, propane, and butane.

Each alkane has a name. The rules for naming all organic compounds are approved by the International Union of Pure and Applied Chemistry (IUPAC). The prefix in the name of the alkane refers to the number of carbons hooked to one another in the chain. The prefixes of the first ten alkanes are as follows:

Prefix	Number of carbon atoms
meth-	1 carbon (no chain possible)
eth-	2
prop-	3
but-	4
penta-	5
hexa-	6
hepta-	7
octa-	8
nona-	9
deca-	10

The suffix -ane indicates that the compound is an alkane with single bonds between the carbons in the chain of carbons. Using the formula $C_nH_{(2n+2)}$, one can translate between names and formulas. For example, methane can be determined to have one carbon atom (meth-) that is surrounded by four hydrogen atoms since ($C_1H_{(2(1)+2)}$) yields four hydrogens. Therefore, the formula of methane is CH_4. Remember, the number of atoms of a particular element in a molecular formula is indicated by a subscript after the symbol for that element.

19

You should be able to recognize the names and molecular formulas of the first 10 alkanes and to translate from name to formula and vice versa.

Name	How the formula is determined	Formula
Methane	meth- = 1 C and $(C_1 H_{(2(1)+2)}) = 4$ H	CH_4
Ethane	eth- = 2 C and $(C_2 H_{(2(2)+2)}) = 6$ H	C_2H_6
Propane	prop- = 3 C and $(C_3 H_{(2(3)+2)}) = 8$ H	C_3H_8

EXERCISE
2·1

Following the same pattern, complete this table for the next seven alkanes.

NAME	HOW THE FORMULA IS DETERMINED	FORMULA
1. butane	_____	_____
2. pentane	_____	_____
3. hexane	_____	_____
4. heptane	_____	_____
5. octane	_____	_____
6. nonane	_____	_____
7. decane	_____	_____

Going in the other direction, from the formula to the name, yields

Formula	How the name is determined	Name
CH_4	1 C = meth- and 4 H is $2n + 2$ $(C_1 H_{(2(1)+2)}) =$ -ane	methane
C_2H_6	2 C = eth- and 6 H is $2n + 2$ $(C_2 H_{(2(2)+2)}) =$ -ane	ethane
C_3H_8	3C = prop- and 8 H is $2n + 2$ $(C_3 H_{(2(3)+2)}) =$ -ane	propane

EXERCISE
2·2

Following the same pattern, complete this table for the next seven alkanes.

FORMULA	HOW THE NAME IS DETERMINED	NAME
1. C_4H_{10}	_____	_____
2. C_5H_{12}	_____	_____
3. C_6H_{14}	_____	_____

4. C_7H_{16} _____ _____

5. C_8H_{18} _____ _____

6. C_9H_{20} _____ _____

7. $C_{10}H_{22}$ _____ _____

For each of the foregoing compounds, all of the carbon atoms are connected one to another by single bonds, which we indicate by a dash when drawing a picture of the structure. Remember that Lewis structures will have the number of valence electrons from each atom included. Carbon has four valence electrons and hydrogen has one.

$$\cdot\dot{C}\cdot \quad H\cdot$$

In most compounds, carbon obeys the octet rule, which indicates that there will be eight shared electrons around the carbon. This is usually accomplished by having four bonds to each carbon atom, with each bond sharing two electrons, for a total of eight electrons; hence the octet. A hydrogen atom with one valence electron does not follow the octet rule and can be bonded to a carbon atom only by a single bond, for a total of two electrons. Methane has a total of eight electrons in its Lewis structure: four from the carbon atom and four from the hydrogen atoms, one per hydrogen. Thus in methane, with only one carbon atom, there must be four single bonds to the four hydrogen atoms. Since these spread out evenly in three-dimensional space, methane has a tetrahedral shape.

$$
\begin{array}{ccc}
 & & H \\
 & & | \\
H\!:\!\ddot{C}\!:\!H & & H-C-H \\
 & & | \\
 & & H
\end{array}
$$

When a Lewis structure is written on paper, the alkane molecules appear to be flat, but molecules are actually three-dimensional, so the geometry is tetrahedral around each of the carbons in an alkane. A close look at the picture of the structure reveals that one carbon is behind and one carbon is more forward. These are indicated in a Lewis structure with

and

In the case of ethane, C_2H_6, with the Lewis structure

$$\begin{array}{ccccc} & H & & H & \\ & | & & | & \\ H & - & C & - & C & - & H \\ & | & & | & \\ & H & & H & \end{array}$$

the geometry is still tetrahedral around each carbon.

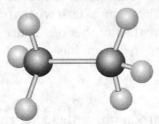

Clearly, the structure of any organic compound can be represented in multiple ways, including Lewis structures and three-dimensional geometric structures, and these in turn are often condensed into structural formulas and line drawings.

Another way to write the formula is to write what is around each carbon in the chain, such that ethane

$$\begin{array}{ccccc} & H & & H & \\ & | & & | & \\ H & - & C & - & C & - & H \\ & | & & | & \\ & H & & H & \\ & CH_3 & & CH_3 & \end{array}$$

is represented as CH_3CH_3. This is called a condensed structural formula. With more than two carbons in the chain, CH_2 since two of the four bonds of central carbons are the carbon-to-carbon bonds, the nonterminal carbons are leaving only two for hydrogen. For example, propane has the Lewis structure

$$\begin{array}{ccccccc} & H & & H & & H & \\ & | & & | & & | & \\ H & - & C & - & C & - & C & - & H \\ & | & & | & & | & \\ & H & & H & & H & \end{array}$$

The condensed structural formula of propane is $CH_3CH_2CH_3$.

For each of the following alkanes, draw the Lewis structure and then write the condensed structural formula.

LEWIS STRUCTURE   CONDENSED STRUCTURAL FORMULA

1. butane   _____   _____

2. pentane   _____   _____

3. hexane   _____   _____

4. heptane   _____   _____

5. octane   _____   _____

In alkanes, each of the two terminal (end) carbons has three hydrogen atoms. In octane there are six central carbons, each with two hydrogen atoms. To condense structural formulas even more, instead of writing

$$CH_3CH_2CH_2CH_2CH_2CH_2CH_2CH_3$$

for Octane we can combine the CH_2 groups inside parentheses, for $CH_3(CH_2)_6CH_2$. Sometimes you might see a "dot," as in $CH_3 \bullet (CH_2)_6 \bullet CH_3$, or a line used between C groups, as in CH_3—$(CH_2)_6$—CH_3, but these formats will not be used in this book.

2·4

Write the condensed formula for each of the following alkanes, combining like CH_2 groups as shown for octane above.

1. butane   _____

2. pentane _____

3. hexane   _____

4. heptane _____

5. nonane   _____

Another condensed way to indicate structure is in a line drawing that provides just the minimum amount of information. To use a line drawing, three carbons are needed in the chain. For an alkane, the carbon connections are shown, but not the hydrogen atoms. Each vertex represents a -CH_2- and the end of a line represents a CH_3. The example of octane above is shown in a line drawing as follows:

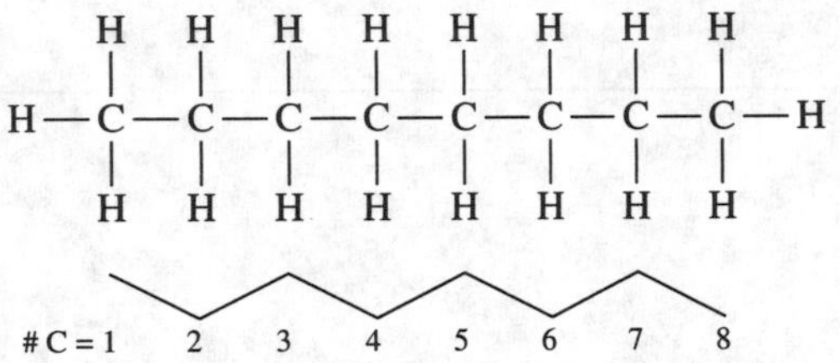

EXERCISE
2·5

Draw the line drawing for each of the following alkanes.

1. heptane _____

2. propane _____

3. butane   _____

4. pentane _____

5. hexane   _____

Sometimes the H atoms are included, as in pentane:

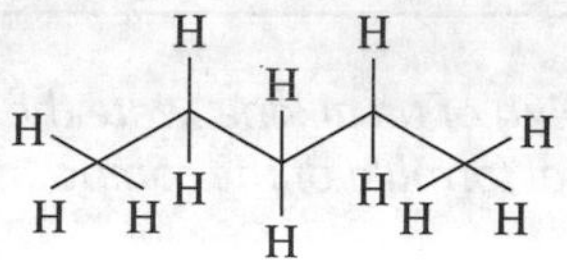

It is important to be able to recognize and visualize a molecule using a variety of different representations, including Lewis structures, molecular geometry, condensed structural formulas, and line drawings.

Using propane as an example, the following model reviews the different ways to draw and write structures.

Propane Lewis structure

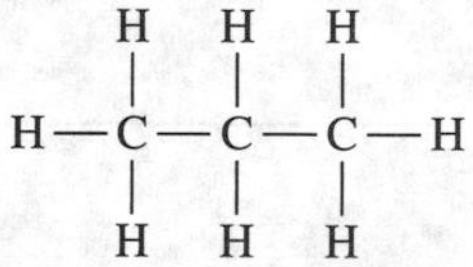

Molecular geometry as represented by the following Ball and Stick Model

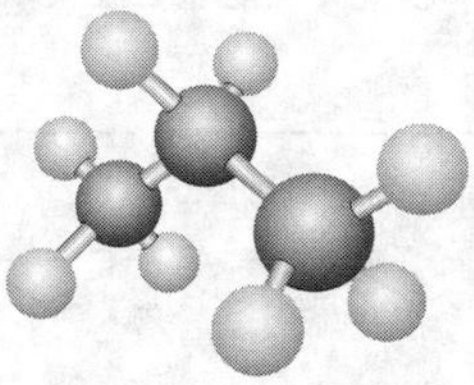

Condensed structural formula

$CH_3CH_2CH_3$

Line drawing

propane

EXERCISE
2·6

Given the molecular formula of an alkane, write the name, the condensed structural formula, and the line drawing of the compound.

1. C_4H_{10} ______________________________

____

2. C_7H_{16} ______________________________

____

3. $C_{10}H_{22}$ ______________________________

____

4. C_5H_{12} ______________________________

____

5. C_6H_{14} ______________________________

____

Given the line drawing of an alkane, write the molecular formula, the name, and the condensed structural formula of the compound.

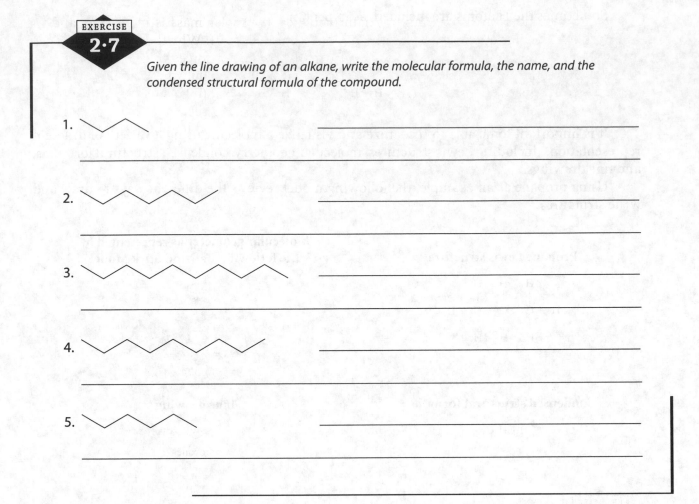

1. _____

   _____

2. _____

   _____

3. _____

   _____

4. _____

   _____

5. _____

   _____

Physical Properties of Alkanes

The physical properties of the alkanes, such as melting point and boiling point, depend on intermolecular forces. Because alkanes contain only carbon and hydrogen, there are no net dipole moments indicating a nonpolar molecule, and the main intermolecular forces present in the liquid and solid forms of alkanes are the London dispersion forces. Since dispersion forces depend on the total number of electrons present, more electrons will be present as more atoms are added to the carbon chain. Hence the longer the chain, the greater the dispersion force. Because all alkanes have the same type of intermolecular force, as more atoms are added to the chain, the molar mass of the compound also increases, as does its boiling point.

A definite pattern is observed in correlating molar mass (MM) with the melting points (m.p.) and the boiling points (b.p.) of alkanes:

Compound	MM	m.p.	b.p.
Methane, CH_4	16	−182.5°C	−161.5°C
Ethane, C_2H_6	30	−183.3°C	−88.6°C
Propane, C_3H_8	42	−189.7°C	−42.1°C
Butane, C_4H_{10}	58	−138.4°C	−0.5°C

As the number of carbons in the chain increases, the molar mass is increasing, and the melting points are also increasing slightly, as are the boiling points. Why the greater increase in boiling points? Remember that when a solid changes to a liquid, the forces holding it in solid form need be overcome only enough to let molecules slide by each other, but when a liquid changes to a gas, the forces between the molecules holding them in the liquid (intermolecular forces) must be totally overcome to separate each molecule from the others as a separated gas molecule. This requires much more energy, resulting in much greater increases in boiling points.

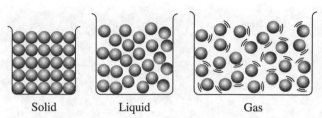

Solid　　　　Liquid　　　　Gas

EXERCISE

2·8

Write the formula of each compound, calculate the corresponding molar mass of each compound, and then predict which liquid has the higher boiling point and explain your reasoning.

1. hexane or octane _____

_____

2. butane or nonane _____

_____

3. methane or decane _____

_____

4. octane or decane _____

_____

5. propane or pentane _____

_____

Naming Simple Branched Alkanes

Not all alkanes are simple, end-to-end long chains of carbon and hydrogen atoms. Branches off the main chain of carbon atoms can form, and such alkanes are called branched, or isomeric, alkanes. Structural isomers have the same chemical formula but different arrangements of the bonds, making different structures. For example, butane and 2-methylpropane (isobutane) have the same formula, C_4H_{10}, but different structures.

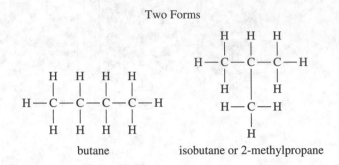

Two Forms

butane   isobutane or 2-methylpropane

However, the different structures must be given different names! To name the branched version, one must first identify the main, or longest, linear chain, and this chain determines the parent name. The group or groups of carbon and hydrogen atoms that are attached to the parent main chain are called alkyl groups. Note that the number of hydrogen atoms in an alkyl group is one less than the number in its alkane name. This is to provide the place of the C-C bonding to the parent chain.

Simple Alkyl Groups

methane   remove a hydrogen →   methyl, $-CH_3$
ethane   remove a hydrogen →   ethyl, $-CH_2CH_3$
propane   remove a hydrogen →   propyl, $-CH_2CH_2CH_3$

To name a branched alkane, first find the longest continuous carbon chain. The longest carbon chain may be right to left, left to right, or even crooked. This is the parent base name of the compound.

After finding the longest chain, then locate branches that are on in the chain. Then number the carbons in the main chain in such a way that the carbons at which the branches occur are located on the lowest possible numbers.

```
                    H
                    |
                H — C — H
        H   H       |       H   H
        |   |       |       |   |
    H — C — C — C — C — C — H
        |   |   |   |   |
        H   H   H   H   H
        1   2   ③   4   5
                   or
        5   4   ③   2   1
```

In this case, numbering in either direction results in the same name.

Name the alkyl groups in alphabetical order, placing the location number in front of the alkyl group. The goal is to write an unambiguous and specific name for the compound.

```
              methyl
        ┌─────────────┐
        │      H      │
        │      |      │
        │  H — C — H  │
        └─────────────┘
        H   H   |       H   H
        |   |   |       |   |
    H — C — C — C — C — C — H
        |   |  ↗|       |   |
        H   H / H   H   H
          Carbon 3
```

The name 3-methylpentane indicates that the methyl is on the third carbon in the chain and that the parent chain is five carbons long.

Next, let's follow the steps to name this structure.

```
                        H
                        |
                    H — C — H
        H   H   H   H   |       H   H
        |   |   |   |   |       |   |
    H — C — C — C — C — C — C — C — H
        |   |   |   |   |   |   |
        H   H   |   H   H   H   H
              H — C — H
                    |
              H — C — H
                    |
                    H
```

Possible chains are as follows:

$$C-C-C-C-C-C-C = 7$$
with $C-C$ branch

The Longest chains.

$$C-C-C-C-C-C-C = 7$$

$$C-C-C-C-C-C-C = 6$$

$$C-C-C-C-C-C-C = 6$$

It is clear from the foregoing structures that there are seven carbons in the longest continuous chain of carbons. There is an ethyl group attached to carbon number 3, and there is a methyl group attached to carbon number 5. This compound is 3-ethyl-5-methylheptane. Number the chain so the alkyl groups are on the lowest numbers. If the numbers are the same both ways, number so the one first alphabetically is on the lower number.

If more than one of the same alkyl group have been added, then a comma is placed between the numbers, and a prefix is placed on the group to indicate the total number. For example, 2,3-dimethylpentane has two methyl groups (the prefix di- on dimethyl means "two"), and they are located on the second and third carbons in the chain. Other common prefixes are tri-, which is the prefix for "three," and tetra-, which is the prefix for "four."

EXERCISE
2·9

Given the structure, write the name for each of the following compounds.

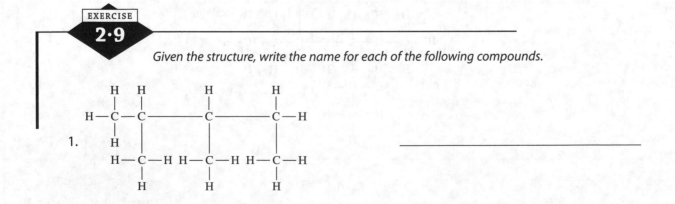

1.

2.

```
        H           H
        |           |
    H — C — H   H — C — H
        |           |       H   H   H   H
        H           H       |   |   |   |
H — C — C —————————— C — C — C — C — C — H
    |   |           |   |   |   |   |
    H   H           H   H   H   H   H
                    H — C — H
                        |
                    H — C — H
                        |
                        H
```                                                    _____

3.

```
        H   H
        | |
    H — C — H   H — C — H
        | |
    H — C — C ————————— C — C — H
        | | | |
        H | | H
    H — C — H   H — C — H
        | |
        H   H
```                                                    _____

4.

```
            H   H
            | |
        H — C — H   H — C — H
    H   H | | H   H
    | | | | | |
H — C — C — C —————————— C — C — C — H
    | | | | | |
    H   H | | H   H
        H — C — H   H — C — H
            | |
            H   H
```                                                    _____

5.

```
                            H
                            |
                        H — C — H
    H   H   H   H   H   H   H   H
    | | | | | | | |
H — C — C — C — C — C — C — C — C — C — H
    | | | | | | | | |
    H   H   H   H   H   H   H   H   H
```                                                    _____

It is also important to practice translating the name of a compound into a structure.

3-ethyl-2,3,4-trimethyloctane

```
        H           H           H
        |           |           |
    H — C — H   H — C — H   H — C — H
        |           |           |   H   H   H   H
        |           |           |   |   |   |   |
H — C — C —————————— C —————————— C — C — C — C — C — H
    |   |           |           |   |   |   |   |
    H   H           |           H   H   H   H   H
                H — C — H
                    |
                H — C — H
                    |
                    H
```

In this name, octane indicates that the base chain has eight carbons; 2,3,4-trimethyl indicates that there are three methyl groups, one each on carbons 2, 3, and 4; and 3-ethyl indicates that there is one ethyl group on the third carbon.

EXERCISE

2·10

Given the name of the compound, draw the corresponding structure.

1. 2,3-dimethylnonane   _____

2. 2,2-dimethyl-4-propylnonane   _____

3. 3-ethyl-4,4-dimethyloctane   _____

4. 2,2,4-trimethylhexane   _____

5. 4-ethyl-2,3-dimethyloctane   _____

Writing the formula of compounds with alkyl groups must also be mastered. Consider the example of 3-ethyl-2,3,4-trimethyloctane and its structure:

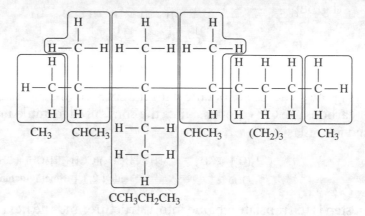

CH_3 $CHCH_3$ $H-C-H$ $CHCH_3$ $(CH_2)_3$ CH_3

$CCH_3CH_2CH_3$

The formula $C_{13}H_{28}$ does not indicate how the molecule is put together and is therefore still written showing how the structure is built:

$$CH_3(CHCH_3)(CCH_3CH_2CH_3)(CHCH_3)(CH_2)_3CH_3$$

EXERCISE
2·11

For each of the compounds in Exercise 2-10, write the corresponding formula.

1. 2,3-dimethylnonane _____

2. 2,2-dimethyl-4-propylnonane _____

3. 3-ethyl-4,4-dimethyloctane _____

4. 2,2,4-trimethylhexane _____

5. 4-ethyl-2,3-dimethyloctane _____

To review, the addition of an alkyl group changes the condensed formula of the base alkane. This is best observed by looking at the structures of the base alkane and the branched chain.

$$
\begin{array}{ccccccc}
& H & H & H & H & H & H & H \\
& | & | & | & | & | & | & | \\
H- & C- & C- & C- & C- & C- & C- & C-H \\
& | & | & | & | & | & | & | \\
& H & H & H & H & H & H & H
\end{array}
$$

Comparing the structure of octane with that of 3-ethyl-2,3,4-trimethyloctane, note that not all the center carbons are CH_2 groups. Since the condensed formula needs to follow the structure, the condensed formulas are

$$CH_3(CH_2)_6CH_3 \qquad CH_3(CHCH_3)(CCH_3CH_2CH_3)(CH_2)_3CH_3$$

Octane   3-ethyl-2,3,4-trimethyloctane

The first step is to translate a name into a structure, and the next step is to use that structure to write the condensed formula.

EXERCISE

2·12

Write the condensed formula for each of the following compounds. (You drew these structures in Exercise 2-10.)

1. 2,3-dimethylnonane   _____

2. 2,2-dimethyl-4-propylnonane   _____

3. 3-ethyl-4,4-dimethyloctane   _____

4. 2,2,4-trimethylhexane   _____

5. 4-ethyl-2,3-dimethyloctane   _____

The addition of an alkyl group also changes the line drawing of a compound. Each alkyl group is placed on the designated carbon in the drawing.

Again compare the structures of octane above and 3-ethyl-2,3,4-trimethyloctane below.

$$
\begin{array}{c}
\text{structure of 3-ethyl-2,3,4-trimethyloctane}
\end{array}
$$

The line drawing of octane is simply lines because there are no additions to the chain. On 2,3,4-trimethyl-3-ethyloctane, the added group is shown off the vertex of the indicated carbon.

EXERCISE
2·13

Draw the line drawing for each compound.

1. 3-methylhexane _____

2. 3,4,5-trimethyloctane _____

3. 3,4-dimethylheptane _____

4. 4-ethyl-2,3-dimethyloctane _____

5. 3-ethyl-4,4-dimethyloctane _____

Isomers have been previously mentioned. Remember isomers have the same (iso- means "same") atomic composition. An isomer of butane was named and identified as an isomer previously. Pentane has three isomers, and hexane has five. The number of isomers increases as the number of carbons increases. Decane, with ten carbons in the chain, has seventy-five possible isomers.

Draw the three possible isomers of pentane, and name them.

Alkanes are not always in open-ended straight chains. Sometimes the ends of the chains hook together. When this occurs, the structure is called a ring, even though it is not perfectly round.

Ring alkanes are indicated by the prefix cyclo-. Simple ones have just the carbon chain name, but the formula has only CH_2 groups. This is because no CH_3 groups can exist in the main ring since the ends are hooked together and there are no terminal carbons where three hydrogens can bond.

Note that in propane, each terminal (end) carbon has three hydrogens and the formula is C_3H_8, whereas cyclopropane has only two hydrogens on each carbon and the formula is C_3H_6.

EXERCISE
2·15

Draw the structures and write the formulas of the following compounds.

DRAWING   FORMULA

1. cyclobutane

2. cyclopentane

_____   _____

3. cyclohexane

_____   _____

4. cycloheptane

_____   _____

5. cycloocatane

_____   _____

Cycloalkanes can have attached alkyl groups too. Since the carbons are in a ring, the numbering of the carbons starts with the added group. For instance, no matter how methyl cyclopentane is drawn, the numbering of the carbons starts with the carbon that has the added methyl group.

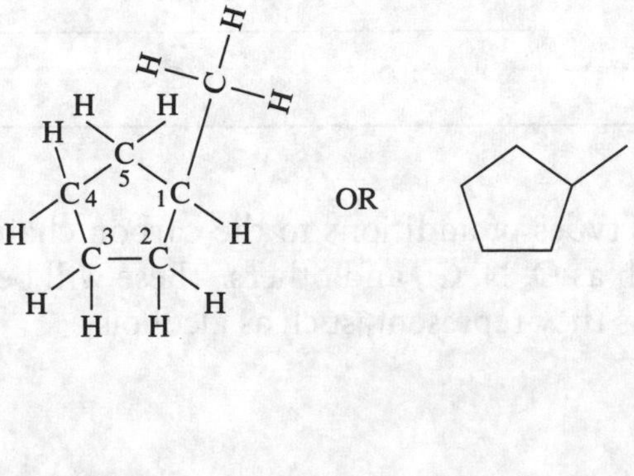

OR

If more than one group has been added to the ring, number the carbons in the ring in such a way that the added groups are on the lowest-numbered carbons.

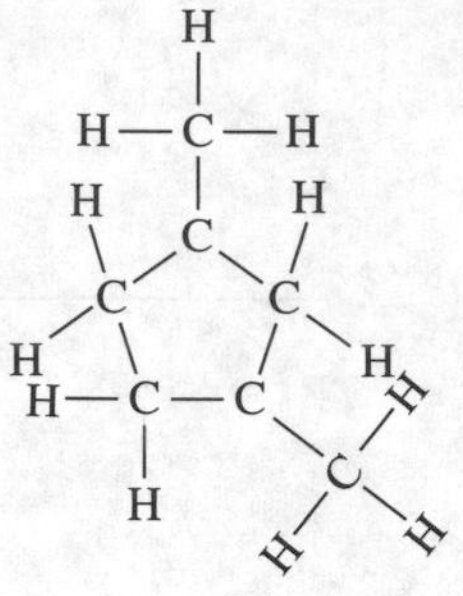

OR

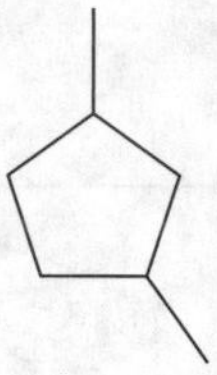

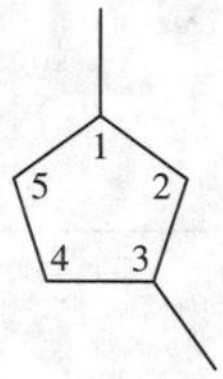

OR

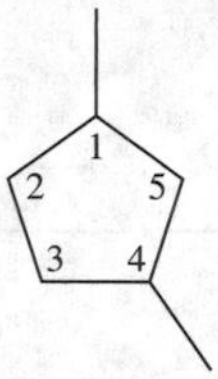

The lowest numbers are 1 and 3
1,3-dimethylcyclopentane.

This could be numbered either so that the methyl groups are on carbons 1 and 3 or so that they are on carbons 1 and 4. The lowest numbers are 1 and 3, so the name of the compound is 1,3-dimethylcylcopentane.

EXERCISE
2·16

Write the names of the following cycloalkanes.

1. _____

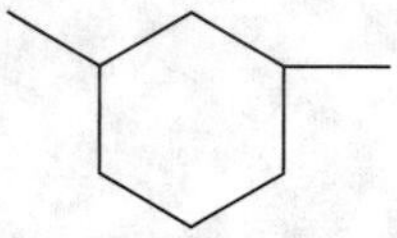

2. _____

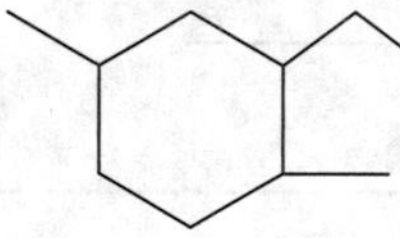

3. _____

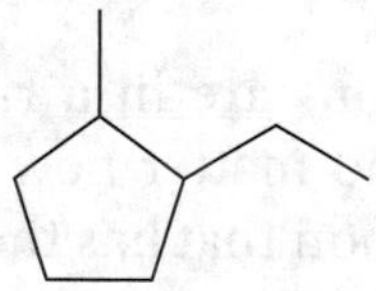

4. _____

5. _____

There are other types of additions to the carbon chain that involve different atoms called "hetero-" atoms, such as O, N, Cl, and others. These will be covered in separate chapters on the groups of compounds they represent, such as alcohols.

Combined Practice

Name the following alkane compounds and write their condensed formulas.

	NAME	CONDENSED FORMULA
1.	_____	_____
2.	_____	_____
3.	_____	_____
4.	_____	_____
5.	_____	_____
6.	_____	_____
7.	_____	_____

8.
```
            H
            |
        H—C—H
   H    H        H
   |    |        |
H—C———C———C———C—H
   |    |        |
   H    H        H
            H—C—H
            |
            H—C—H
            |
            H
```
_____   _____

9.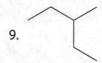

_____   _____

10.
```
        H              H
        |              |
    H—C—H          H—C—H
   H    H    H    H    H
   |    |    |    |    |
H—C——C——C——C——C——C——C—H
   |    |    |    |    |    |
   H    H    H    H    H    H
            H—C—H
            |
            H
```
_____   _____

EXERCISE
2·18

Draw the Lewis structures and line drawings of the following compounds.

LEWIS STRUCTURE   LINE DRAWING

1. 2,4,4-trimethylhexane

_____   _____

2. 4-propyl-3-methylheptane

_____   _____

3. 1,2-dimethylcyclohexane

_____   _____

4. 3-ethyl-2,3-dimethyloctane

_____   _____

5. 2-ethyl-2-methylheptane

_____   _____

6. 3-methylhexane

_____   _____

7. 3,4,5-trimethlyoctane

_____   _____

8. 3,3-diethylpentane

_____   _____

9. 1-ethyl-2-methylcyclobutane

_____   _____

10. 1-methyl-2-propylcyclopentane

_____   _____

Alkenes

It is helpful to be able to classify organic compounds according to similar features in their structure. In organic chemistry, formulas alone often do not give enough information for us to determine the structures of organic molecules. In the previous chapter, we studied a class of organic saturated hydrocarbon compounds called alkanes that have carbon-carbon single bonds in the main carbon chain, and we found that the length of the carbon chain is used to help name organic molecules.

How is an alkene different from an alkane? Alkane chains have only single bonds between the carbon atoms, whereas when the main carbon chain has one or more carbon-carbon double bonds, the compound is classified as an alkene. If there is only a single double bond, this results in alkenes having the general formula C_nH_{2n}. If there are two carbons, then $n = 2$ and the number of hydrogen atoms is 2×2, or 4. The formula is therefore C_2H_4. Alkenes are unsaturated because a double bond can be broken to add more hydrogen atoms to the compound.

unsaturated

saturated

propene

propane

Double bond can be broken to add more hydrogen atoms

Maximum number of hydrogen atoms bonded

As with alkanes, the main name of the alkenes is based on the number of carbons in the longest carbon chain. A number identifying the location of the double bond is used, and then the prefix associated with the number of carbons in the longest chain is included in the name. This is followed by the suffix -ene to indicate the double bond. Just as with alkanes, we must number the carbons in the longest chain from both directions, but in alkenes the direction is chosen to indicate the location of the carbon the double bond. The number used in front of the name is from the direction providing the lowest possible number for the double bond. In alkanes the lowest number of an alkyl group is considered the most important, but in alkenes the double bond getting the lowest number takes precedence. The steps used to name the base organic compound are as follows

- Identify the longest chain and write its prefix.
- What types of bonds are present in the chain? Use the appropriate suffix.

◆ If a multiple bond is present, identify what carbon in the chain the multiple bond starts on. If the compound is a ring structure (cyclo-), the carbons are numbered so the double bond is between the first carbon and the second carbon.

For example, let us use these steps to identify the following Lewis structure.

$$H_2C=CH-CH_2-CH_3$$

In this case the longest chain is four carbons with a prefix of but-, and a double bond is present, indicating that the suffix is -ene. The main name is butene. Numbering from left to right the double bond is on the first carbon, and numbering from right to left the double bond starts on the third carbon. The number used in front is the lower number, 1. The correct name is 1-butene.

As another example, C_3H_6 can be represented with the following Lewis structure:

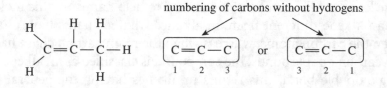

In the Lewis structure, the longest continuous carbon chain has three carbons. The prefix for the base name is prop-. Since the compound has one double bond, the suffix to use is -ene. Thus the base name of the compound is propene. Since the double bond can be between any two carbons in the longest carbon chain, we need to identify where the double bond is located. Numbering from left to right the double bond is on the first carbon, and numbering from right to left it is located on the second carbon. The number "1" is used to indicate that the double bond starts with the first carbon. The number "2" is used if it starts on the second carbon, and so on. The lowest number is "1," so the name is 1-propene. Remember that you must count from both ends and that the multiple bond needs to be on the lowest-numbered carbon possible. Using this rule on propene will always yield the number "1" and is therefore just called propene.

An exception to this numbering is ethene. Ethene has only two carbons in the chain, so the only place the carbon can be is between the first carbon and the second carbon, no matter what direction it is numbered from. Therefore, a number does not have to be included for ethene.

$$H_2C=CH_2$$
ethene

numbering of carbons without hydrogens

$$\boxed{C=C}$$ or $$\boxed{C=C}$$
  1   2   2   1

If there is more than one multiple bond, a prefix is placed in the name before the -ene to indicate the number of multiple bonds. The suffix -diene indicates that there are two double bonds in the compound. Thus 1,3-butadiene has a base chain of four carbons and two double bonds: one between the first and second carbons, and one between the third and fourth carbons.

$$H_2C=CH-CH=CH_2$$

four carbons

two double bonds, on the first and third carbons

If alkyl groups are also added to the base chain, again they are indicated in alphabetical order (ethyl- before methyl-) and the numbers indicating their location are added. Consider

The longest chain is six carbons and there is a double bond, for a base name of hexene. The numbering is done in such a way as to locate the double bond on the lowest-numbered carbon first.

Either way it is numbered, the double bond is on the third carbon. In a case such as this, where the carbon number will be the same, choose the direction where addition substituents are on the lowest-numbered carbon.

Thus, numbering then from left to right, the methyl group is located on the second carbon, resulting in the name 2-methyl-3-hexene.

Ring structures also exist with alkenes. The numbering of the carbons always begins with the location of a double bond. If alkyl groups are added, the double bond still takes precedence in the numbering.

The ring has six carbons and a double bond, for a base name of cyclohexene. Numbering clockwise from the double bond, there is a methyl group on carbon 1 and on carbon 2. The name of this compound is 1,2-dimethylcyclohexene.

If more than one double bond is present in the ring, both are indicated in the name by a location number, and the suffix is changed appropriately for the number of double bonds—that is, -diene for two double bonds, -triene for three double bonds, and -tetraene for four double bonds.

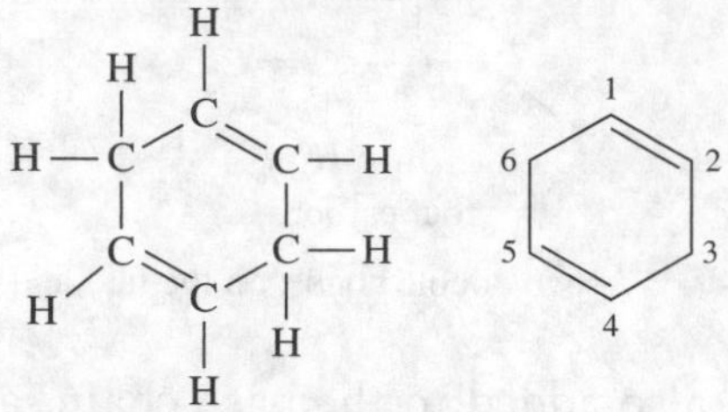

There are six carbons in the chain, and the chain is in the form of a ring, indicating the prefix cyclohexa-. There are two double bonds, indicating the suffix -diene. This yields the name cyclohexadiene. To indicate the placement of the double bonds, numbering clockwise from the first location, the numbers are 1 and 4. The complete name is 1,4-cyclohexadiene.

EXERCISE
3·1

Determine the name of each of the following organic structures.

1. 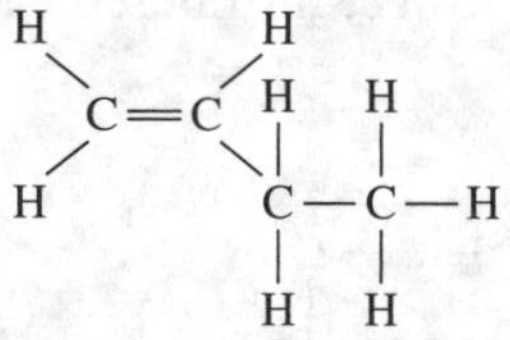 or _____

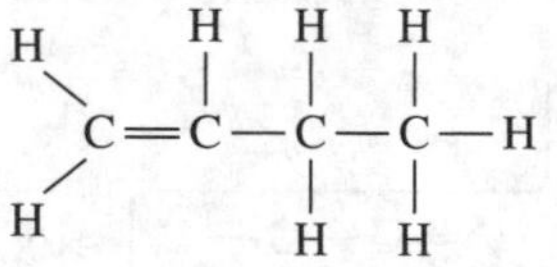

2. 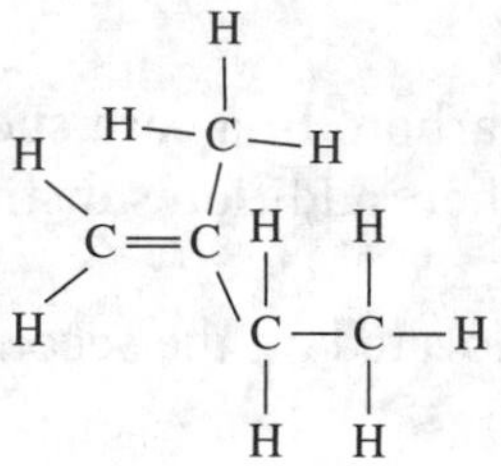 or _____

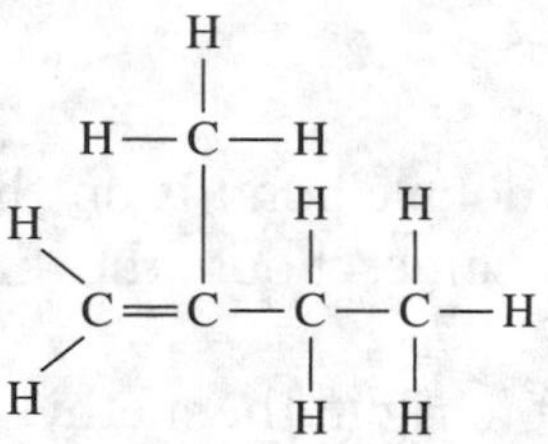

3. _____

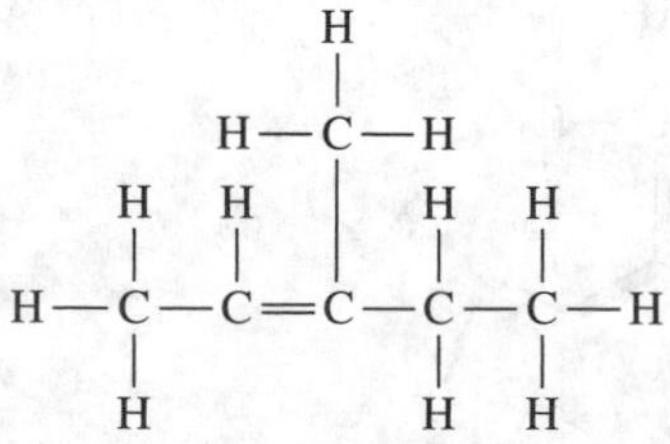

4.

```
            H
            |
        H — C — H
            |
  H   H   H   H   H   H
  |   |   |   |   |   |
H—C — C — C = C — C — C—H
  |   |       |   |   |
  H   H       H   H   H
            |
        H — C — H
            |
            H
```
_____

5.

```
            H
            |
        H — C — H
            |
  H   H   H   H   H   H       H
  |   |   |   |   |   |      /
H—C — C — C — C — C — C = C
  |   |   |   |   |          \
  H   H   H   |   H   H        H
            H—C—H   H—C—H
              |       |
            H—C—H   H—C—H
              |       |
              H       H
```
_____

6.

```
      H       H
      |       |
  H — C — H H — C — H
      |       |
  H   H   H   H   H   H
  |   |   |   |   |   |
H—C — C — C = C — C — C—H
  |   |               |   |
  H   |               H   H
    H—C—H
      |
      H
```
_____

7.

```
          H
          |
      H   C   H
       \ / \ /
    H—C     C—H
       |     |
    H—C —— C—H
       |     |
       H     H
```
_____

8.

```
          H
          |
      H — C — H
          |
  H   H   H   H       H
  |   |   |   |      /
H—C — C — C — C = C — C—H
  |   |   |          |   |
  H   H   |          H   H
        H—C—H   H—C—H
          |       |
        H—C—H     H
          |
          H
```
_____

9.

```
      H   H   H       H
      |   |   |      /
  H—C—C—C=C
      |   |   |      \
      H   |   H       H
          H—C—H
              |
              H
```

_____

10.

```
  H       H   H       H
   \      |   |      /
    C=C—C—C=C
   /      |   |      \
  H       H   H       H
```

_____

Reversing this process and determining the base structure from the name requires dissecting the name.

- What base term is used to indicate the longest chain?
- What suffix is used to indicate the types of bonds used?
  - If a multiple bond(s) is(are) present, what number(s) is(are) used to indicate its(their) location(s)?
- Are any alkyl groups added in front of the base name?
  - If an alkyl group(s) is(are) added, what number(s) is(are) used to indicate its(their) location(s)?
- Fill the structure with hydrogen atoms so that all carbons have four bonds.

2-Methyl-1,3-butadiene would be broken down as follows:

The prefix buta- indicates that the longest chain is four carbons long.

$$C—C—C—C$$

The suffix -diene indicates two double bonds, and -1,3- indicates that they are on the first and third carbons.

$$C=C—C=C$$

Next, 2-methyl indicates that a CH_3 group is added on the second carbon.

$$
\begin{array}{c}
H \\
| \\
H—C—H \\
| \\
C=C—C=C
\end{array}
$$

The last step is to make sure that all carbons have eight valence electrons in four bonds. If they need additional bonds, hydrogen is bonded to the carbon. Looking at the structure reveals that the first carbon has a double bond and needs two additional bonds with hydrogen, the second carbon has four bonds and needs no additional bonds, the third carbon has three bonds and needs one additional bond with hydrogen, and the fourth carbon has two bonds and needs two additional bonds with hydrogen.

Note that each carbon still obeys the octet rule since each carbon has four bonds utilizing eight valence electrons.

EXERCISE
3·2

Draw the organic structure for each of the following organic compounds.

1. 3-hexene

2. 2,3-dimethyl-2-butene

3. 3-methyl-2-pentene

4. 1,4-cyclohexadiene

5. 1,3-dimethylcyclopentene

_____

6. 2,3,3-trimethyl-1,4,6-octatriene

_____

7. 3-ethyl-2,2-dimethyl-3-heptene

_____

8. 2-methyl-1,3-butadiene

_____

9. 2,3-dimethylcyclopentene

_____

Just like alkanes, alkenes can also be represented by a condensed structural formula or a line drawing. The difference is that in alkenes, a double bond must be indicated.

From the name of propene, first draw the structural formula of propene. The condensed structural formula can then be determined. The double bond is indicated by double lines.

EXERCISE
3·3

For each of the following compounds, which were drawn in Exercise 3-2, write the condensed structural formula.

1. 3-hexene

_____

2. 2,3-dimethyl-2-butene

_____

3. 3-methyl-2-pentene

_____

4. 1,4-cyclohexadiene

_____

5. 1,3-dimethylcyclopentene

_____

6. 2,3,3-trimethyl-1,4,6-octatriene

_____

7. 3-ethyl-2,2-dimethyl-3-heptene

_____

8. 2-methyl-1,3-butadiene

_____

9. 2,3-dimethylcyclopentene

_____

From the structure and the condensed structural formula, a line drawing can also be made. Again, note that a double line is used for the double bond.

EXERCISE

3·4

Make a line drawing for each of the compounds drawn in Exercise 3-2

1. 3-hexene

2. 2,3-dimethyl-2-butene

3. 3-methyl-2-pentene

4. 1,4-cyclohexadiene

5. 1,3-dimethylcyclopentene

6. 2,3,3-trimethyl-1,4,6-octatriene

_____

7. 3-ethyl-2,2-dimethyl-3-heptene

_____

8. 2-methyl-1,3-butadiene

_____

9. 2,3-dimethylcyclopentene

_____

In the chapter on alkanes and so far in this chapter, only alkyl groups have been substituted into the structures, but other atoms or groups could have been substituted as well. Some of these are found in the following table.

Some Names of Common Substituent Groups	
Functional group	**Name**
$-NH_2$	amino
$-F$	fluoro
$-Cl$	chloro
$-Br$	bromo
$-I$	iodo
$-NO_2$	nitro
$-CH=CH_2$	vinyl

The same rules that apply when these groups are substituted into alkanes and when they are substituted into alkenes. Consider, for example, the alkane structure

This structure is called 2-amino-3-bromohexane. The base chain is six carbons long, and only single bonds are present, so the base name is hexane. Numbering the carbons to have the lowest numbers, the amino group is on the second carbon and the bromo substituent is on the third carbon. In the name, the substituents are listed alphabetically, so the amino group is listed first.

This alkene example has a ring structure with six carbons (cyclohex) and there are three double bonds (triene), which are located on the first, third, and fifth carbons. Two chlorines (-dichloro have been added, one on the first carbon and one on the fourth carbon. The name of this molecule is 1,4-dichloro-1,3,5-cyclohexatriene.

EXERCISE 3·5

Draw structures for the following compounds.

1. 1-bromo-4,4-dichloro-2-methylcyclohexene _____

2. 2-fluoro-3-nitrocyclopentene _____

Name each of the following compounds. (It may be helpful to draw the structures first to determine the location of double bonds and if the structure is a ring.)

1. $CH_3CHCHCH_2C(Br_2)CH_2CH_3$ _____

2. $CH=CHCH(Cl)CH_2CH(CH_3)CH_2$ _____

The names of organic compounds can get very long. Some of these structures are found so often in organic chemistry that they are given recognized shorter names. An example is 1,3,5-cyclohexatriene, which is regularly called benzene. Benzene is the parent compound of a large group of compounds called aromatic compounds. Note the variety of ways in which benzene can be represented:

	or		or		or	
		carbon atoms are understood to be at each corner		both the backbone and multiple bonds are shown		indicates that double bonds are alternating

With the recognized name of benzene, any alkyl groups added to the structure are indicated by location and in alphabetical order. Some, such as methylbenzene, have other names. Methylbenzene goes by the name toluene.

There is also a recognized abbreviation for certain combinations of substituted groups on benzene. The prefixes are *o- (ortho-)* for a 1,2 configuration, *m- (meta-)* for a 1,3 configuration, and *p- (para-)* for a 1,4 configuration.

o
1,2

m
1,3

p
1,4

o-dichlorobenzene   *m*-dichlorobenzene

p-dichlobenzene

For example, 1,4-dichloro-1,3,5-cyclohexatriene is called 1,4-dichlorobenzene or *p*-dichlorobenzene.

EXERCISE

3·7

Draw the structures of the following compounds.

1. *o*-dibromobenzene

2. *m*-dibromobenzene

3. 1-bromo-3-nitrobenzene

_____

4. 2-choro-3-nitrobenzene

_____

5. *o*-chlorotoluene

_____

Among the possible substituted groups is phenyl. Benzene minus one hydrogen is called a phenyl group. In the following example, the base chain is three carbons long, and a phenyl group has been added on the second carbon. This molecule is called 2-phenylpropane.

Benzene groups can also be fused together in chains, with the two ring groups sharing two carbon atoms. These chains are given names such as naphthalene for two rings fused together, and anthracene for three rings fused together.

benzene   naphthalene   anthracene

Familiarity with the naming of organic compounds is critical in learning organic chemistry.

Molecular Geometry and Isomers

The double bond changes the geometry around the carbons on either side of it. Three electron domains will be present, instead of four, because the double bond sharing four electrons counts as one domain. This will change the angles and structure of the compound. Moving the three domains to the maximum distance from each other results in a trigonal shape around the carbon.

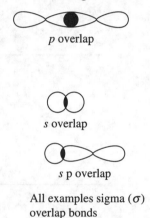

carbon 1
three domains

carbon 2
three domains

single-bonded carbon   three domains at maximum
four domains   distance, for 120° angles

In the chapter on alkanes, all bonds were single bonds. Single bonds consist of the straight overlap of atomic electron orbitals and are called sigma bonds.

p overlap

s overlap

s p overlap

All examples sigma (σ)
overlap bonds

In alkenes, each of the two carbons that make up the double bond have a trigonal planar shape with one pi bond and one sigma bond. The sigma bond consists of a straight-across overlap of atomic orbitals, whereas a pi bond consists of a side-by-side overlap of atomic p-orbitals.

σ

π

side overlap
π bond

Single bonds are always composed of sigma bonds, whereas double bonds are composed of a sigma bond and a pi bond. In ethene there are five sigma bonds and one pi bond:

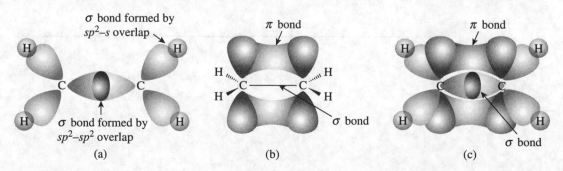

σ bond formed by
sp^2–*s* overlap

σ bond formed by
sp^2–sp^2 overlap

(a)

π bond

σ bond

(b)

π bond

σ bond

(c)

EXERCISE

3·8

Draw each of the following molecules, and indicate the total number of sigma bonds and pi bonds.

1. 1-butene C_4H_8 ____

2. 1-pentene C_5H_{10} ____

3. 1-heptene C_7H_{14} ____

4. 1,4-pentadiene ____

5. 2-methyl-1,3-butadiene ____

If the carbon-carbon double bond is between the second and third carbons in the chain, the number "2" is used to identify where the double bond starts. Because of the trigonal shape around the double bond on the carbons in the double bond, another characteristic is the way the chain continues after the double bond. Look at the following two diagrams:

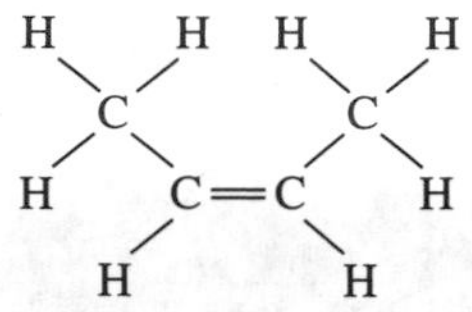

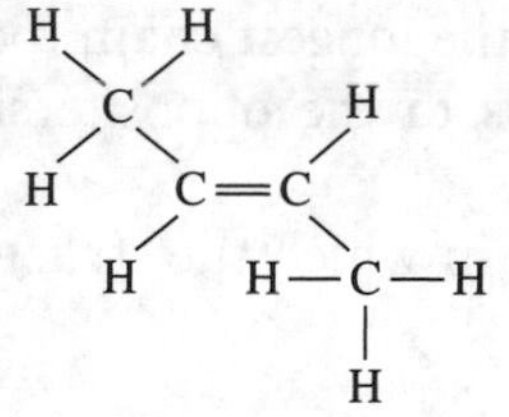

These structures have the same basic formula. Remember that compounds with the same formula but different structures are called structural isomers. The properties of structural isomers are different.

The isomers in both of these diagrams have four carbons in the chain (but-), and both have a double bond between the second and third carbons (-ene), so both represent 2-butene. To indicate the difference between the two diagrams, *cis-* and *trans-* are used. In the first diagram, the chain continues on the same, "up" side as the chain leading into the double bond, whereas in the second diagram, the chain continues on the opposite, "down" side.

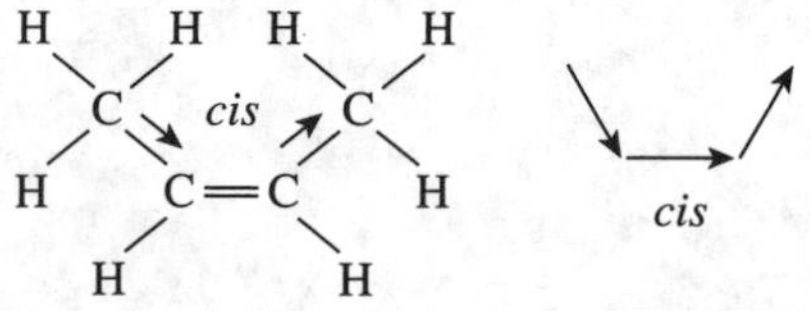

cis-2-butene

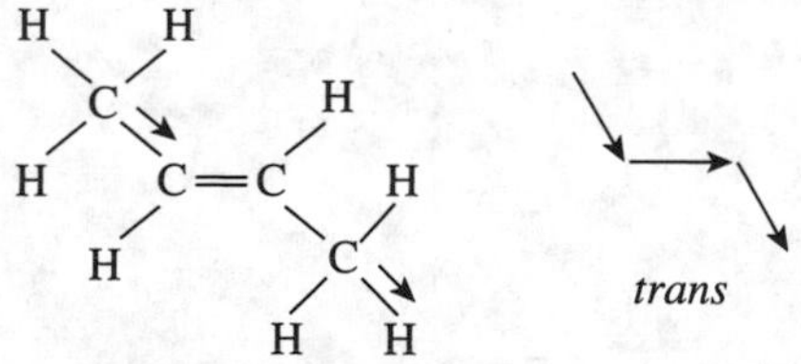

trans-2-butene

In this notation, then, *cis-* is used when the attached groups are on the same horizontal side, and *trans-* is used when the chain continues to the opposite side across the double bond. Pentene offers another example.

cis-2-pentene C$_5$H$_{10}$

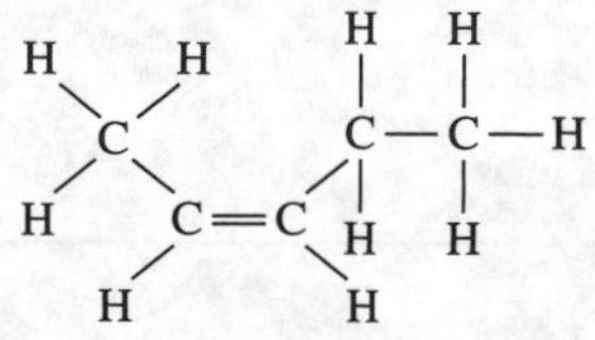

There are five carbons in the longest chain (pent-) with a double bond (-ene) on the second carbon. The chain continues up to the same side (*cis-*). The name of this compound is *cis*-2-pentene.

trans-2-pentene C_5H_{10}

There are five carbons in the longest chain (pent-) with a double bond (-ene) on the second carbon. The chain goes across to the opposite side (*trans-*). The name of this compound is *trans*-2-pentene.

This is an additional way in which the structure of the compound affects the name of the compound in alkenes.

EXERCISE
3·9

Given the molecular formula of an alkene, draw the structure of the compound and name the compound.

1. C_4H_8 The carbon double bond starts with carbon 2, and the chain continues downward.

2. C_7H_{14} The carbon double bond starts with carbon 3, and the chain continues upward.

3. $C_{10}H_{20}$ The carbon double bond starts with carbon 4, and the chain continues downward.

4. C$_5$H$_{10}$ The carbon double bond starts with carbon 2, there is a methyl group attached to carbon 2 (downward), and the chain continues downward.

5. C$_6$H$_{12}$ The carbon double bond starts with carbon 1, there are methyl groups attached to carbon 1 (downward) and carbon 2 (upward), and the chain continues downward.

EXERCISE

3·10

Given the name of an alkene, draw the Lewis diagram.

1. 3-hexene

2. 3-heptene

3. 4-nonene

_____

4. 1,3-octadiene

_____

5. 3-methyl-2-heptene

_____

6. 4-methyl-*cis*-2-hexene

_____

EXERCISE
3·11

Given the name of an alkene, draw the condensed structural formula.

1. 3-hexene

2. 3-heptene

3. 4-nonene

4. 1,3-octadiene

5. 3-methyl-2-heptene

6. 4-methyl-*cis*-2-hexene

EXERCISE

3·12

Write the molecular formula of each of the following compounds.

1. $CH_3CH_2CH{=}CHCH_3$ ____________________

2. 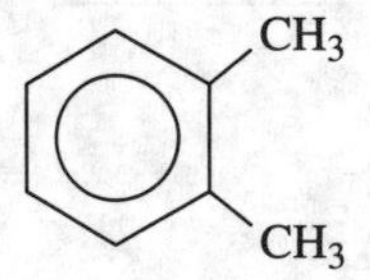

____

3. 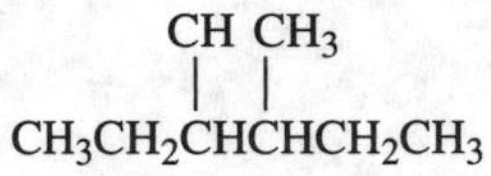

____

4. 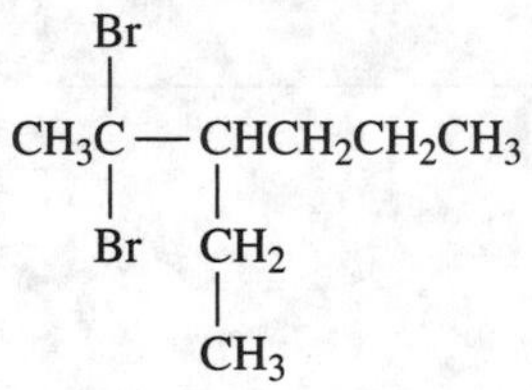

____

5. $CH_3CH_2CH{=}CH_2$ ____________________

Write the name of each of the following compounds.

1.

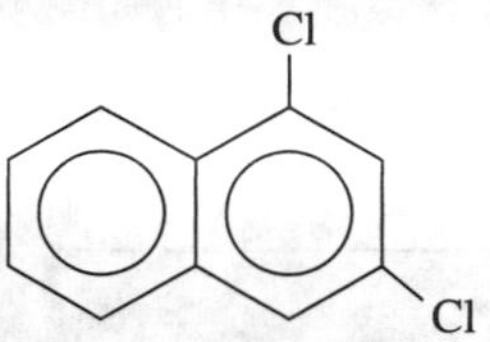

2. 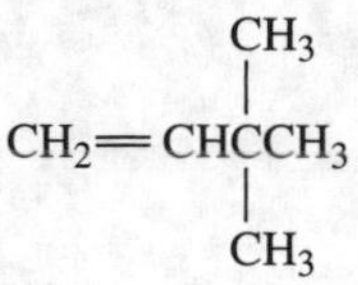 $CH_2\!=\!CHCCH_3$ with CH_3 above and CH_3 below

3. $CH_3C\!=\!CHCH_3$ with CH_3 below

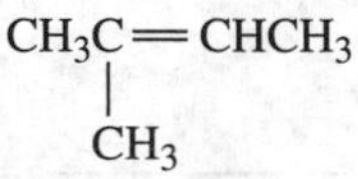

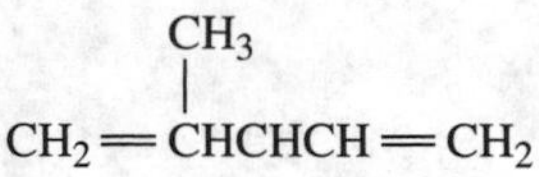

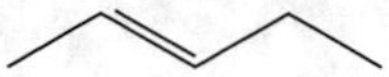

4. $CH_2\!=\!CHCHCH\!=\!CH_2$ with CH_3 above

5.

Draw the Lewis diagram for each of the following compounds.

1. 3-heptene

2. cyclopentene

3. 1,3-butadiene

_____

4. 1-ethyl-2-methylbenzene

_____

5. 2,4-dimethyl-2-pentene

_____

Make a line drawing for each of the following compounds.

1. 1,4-dibromobenzene

_____

2. 2-bromo-3-methyl-2-butene

_____

3. *meta*-dimethylbenzene

_____

4. 1,2-pentadiene

_____

5. 1,3-cyclopentadiene

Given the line drawing of an alkene, write the molecular formula, the name of the compound, and the condensed structural formula.

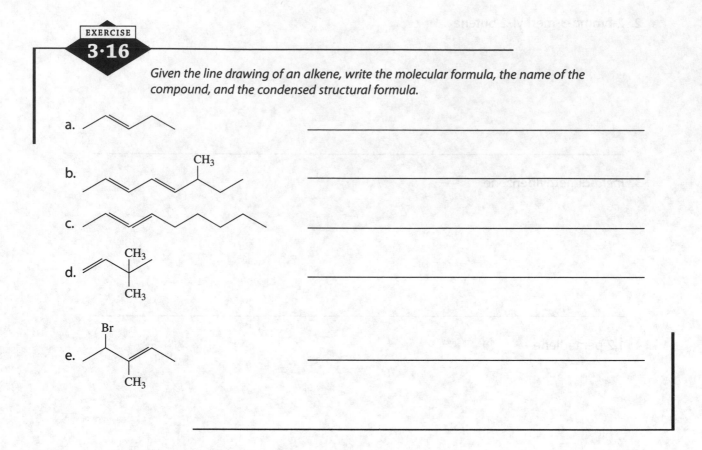

a. _____

b. _____

c. _____

d. _____

e. _____

As with alkanes, functional groups can be added to the chains in alkenes. The names of the compounds will change in accordance with the groups added.

Physical Properties of Alkenes

The physical properties of the alkenes, such as melting point and boiling point, depend on intermolecular forces. Because hydrocarbon organic molecules contain only carbon and hydrogen, there are no net dipole moments, and the main intermolecular forces present in the liquid and solid form of alkenes are London dispersion forces.

The London dispersion force increases with increasing numbers of electrons, so C_4H_8 has greater dispersion forces than either C_2H_2 or C_3H_6 and thus has a higher boiling point as well. When other groups are substituted, other forces (such as dipole-dipole forces and hydrogen bonding) may also affect the properties of the compound. When other groups are added, drawing the structure is critical in determining whether dipoles cancel each other. When comparing alkanes to alkenes of similar molar mass, the alkene has a greater London IMF due to the pi bond in the alkene.

Identify which member of each of the following pairs of hydrocarbon compounds has the higher boiling point.

1. 1. 3-hexene or hexane _____

2. 3-heptene or *trans*-2,3-dimethyl-2-butene _____

3. 4-nonene or 1,4-dimethylbenzene _____

4. 1, 3-octadiene or cyclooctane _____

5. 3 methyl-2-heptene or 3,3,4,4-tetramethylpentene _____

Alkynes

Alkynes are unsaturated organic compounds that have the general formula $C_nH_{(2n-2)}$. Like the alkanes and alkenes, basic alkynes contain only the elements carbon and hydrogen. Hence, alkynes are also classified as hydrocarbon compounds. Like other hydrocarbon compounds, alkynes exhibit only dispersion forces between molecules and therefore are nonpolar and insoluble in water. What makes them different from alkanes and alkenes is the presence of a triple bond.

$$:C:::C: \quad \text{or} \quad :C\equiv C:$$

The triple bond shares six electrons and uses up three of the possible bonds for each carbon in one domain. Each carbon in the triple bond can then have only one additional bond sharing two electrons. Unlike the alkene, which has different *cis-* and *trans-* angles in and out of the bond, the alkyne structure in and out of the triple bond is linear in shape. This is because the maximum distance between two domains is associated with bond angles of 180°.

$$H - C \equiv C - H$$
$$\underbrace{\qquad}_{180°} \underbrace{\qquad}_{180°}$$

The triple bond contains one sigma overlap bond and two pi sideways-overlap bonds.

$$C\equiv C \quad \begin{matrix} \nwarrow \pi \\ \sigma \\ \swarrow \pi \end{matrix}$$

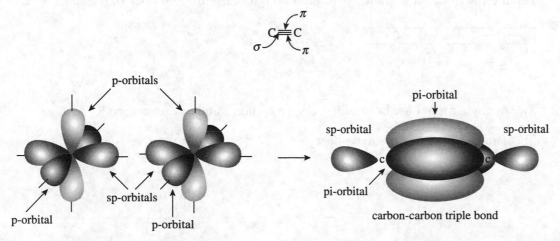

sp hybridized carbon atoms

carbon-carbon triple bond

Triple bonds are stronger than single and double bonds and therefore harder to break. This means in reactions it takes more energy to break the triple bonds in alkynes than the double bonds in alkenes.

As with all organic molecules, there is a systematic way to name the alkynes. Each alkyne has a unique IUPAC name and often may have a common name. As with alkanes and alkenes, the number of carbons in the longest chain is designated with a prefix. The suffix -yne is used to indicate a triple bond, and as in alkenes, the location of the multiple bond is noted with the lowest possible number of carbon in the chain and takes precedence over any other additions to the chain in numbering. If the triple bond begins on the first carbon, the number "1" is used. When substituents are added, the same rules that are used with alkanes and alkenes apply. In the name they appear in alphabetical order with the number of the carbon they occur on given first.

Example 1 Name the following organic structure.

$$H—C≡C—H$$

Two carbons call for the prefix eth- and a triple bond calls for the suffix -yne. Either way the carbons are numbered, no additional groups are added.

The name is 1-ethyne or ethyne since the triple bond can be only on the first carbon.

Example 2 Name the following organic structure.

Seven carbons call for the prefix hept-.

Numbering from left to right, the triple bond is on carbon 3 and calls for the suffix -yne.

Two methyl groups have been added: one on carbon 2 and one on carbon 5.

The name is thus 2,5-dimethyl-3-heptyne.

EXERCISE 4·1

Given the following structures, name each compound and write its formula.

	NAME	FORMULA
1. 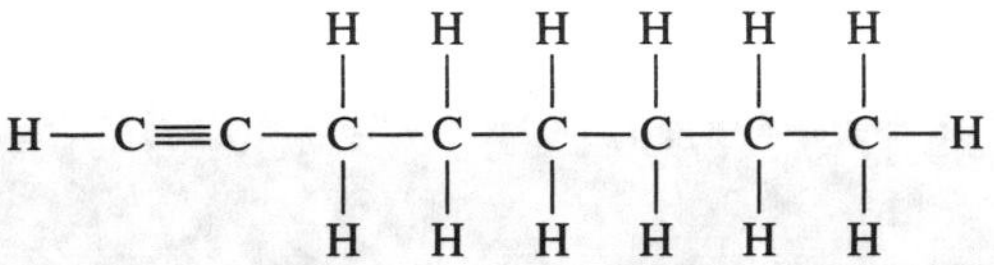	________________	________________
2. 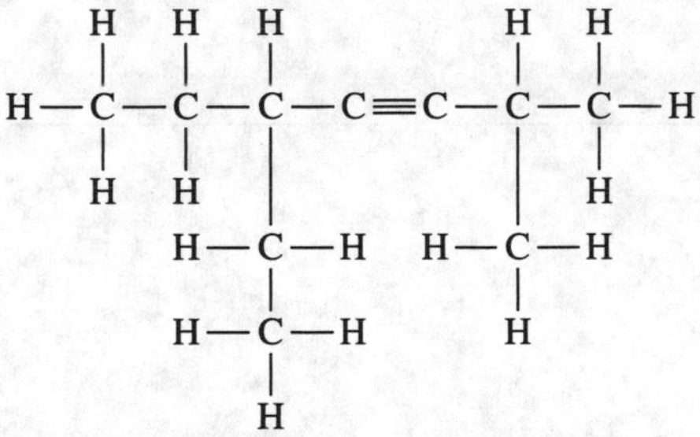	________________	________________
3. 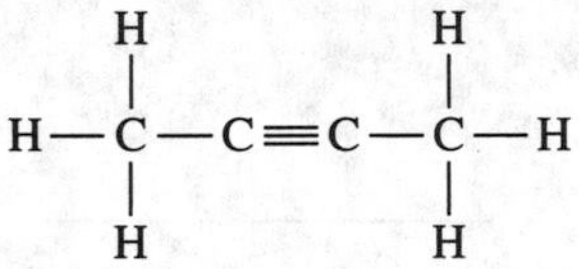	________________	________________
4. 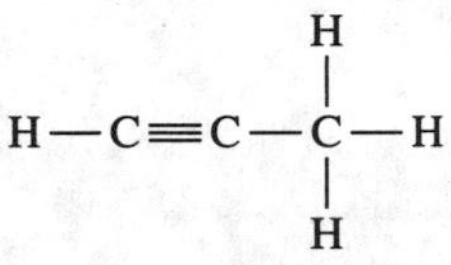	________________	________________
5. 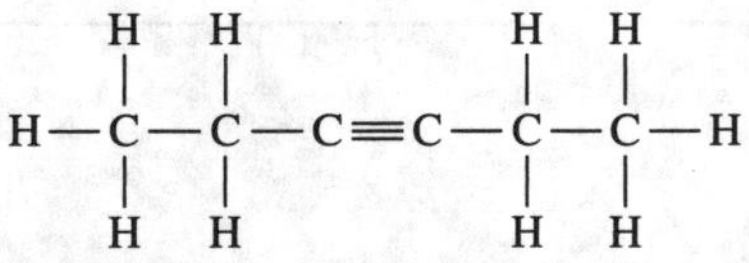	________________	________________
6. 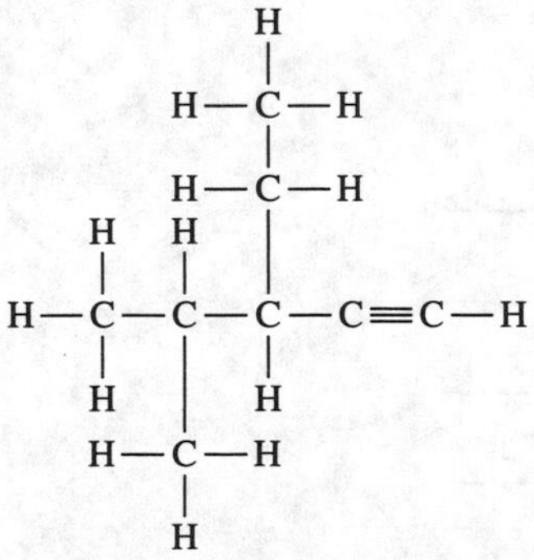	________________	________________
7. 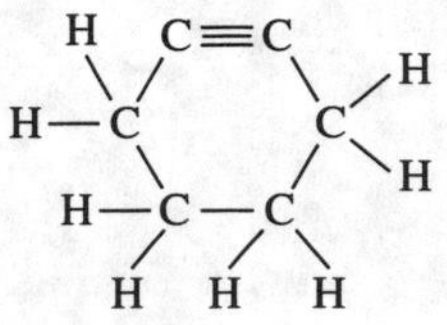	________________	________________

8.

```
                  H
                  |
            H — C — H
                  |
    H             H     H
    |             |     |
H — C — C ≡ C — C — C — C — H
    |             |     |
    H             H     H
                  |
            H — C — H
                  |
                  H
```

_____   _____

9.

```
                      H
                      |
                H — C — H
                      |
    H         H       H
    |         |       |
H — C — C ≡ C — C           C — C — H
    |         |       |     |
    H       H — C — H       H
              |       H — C — H
        H — C — H           |
              |             H
              H
```

_____   _____

10.

```
                      H
                      |
                H — C — H
    H   H             H
    |   |             |
H — C — C — C ≡ C — C — C — H
    |   |             |
    H   H             H
      H — C — H     H — C — H
          |             |
      H — C — H     H — C — H
          |             |
          H             H
```

_____   _____

Given the name of the compound, write its formula.

1. ethyne _____

2. propyne _____

3. 2-butyne _____

4. 1-pentyne _____

5. 1-hexyne _____

6. 3-phenyl-butyne _____

7. 6-methyl-3-octyne _____

8. 2-hexyne _____

9. 3,3-dimethyl-4-octyne _____

10. 4-octyne _____

We can represent the structure of any organic compound in several different ways. It is important to be able to recognize and visualize a molecule using a variety of different representations. These are the same representations that are used with alkanes and alkenes. The difference is the location of the triple bond.

Here are some of the representations for 1-butyne:

Lewis Structure

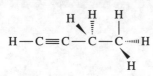

$$H—C≡C—\overset{\displaystyle H}{\underset{\displaystyle H}{C}}—\overset{\displaystyle H}{\underset{\displaystyle H}{C}}—H$$

Molecular Geometry

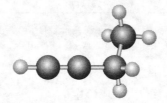

Condensed Structural Formula

CHCCH$_2$CH$_3$

Line Drawing

Remember that a "line drawing" shows just the minimum amount of information. If the molecule is a hydrocarbon, the carbon connections are shown, but not the hydrogens. For 1-butyne, the first vertex represents a -C≡C- triple bond. The next vertex represents

CH$_2$

Triple bond   CH$_3$

-CH$_2$- and the last of a line represents a CH$_3$ group. If the triple bond is in a different position, the triple is still noted in the structure.

If the triple bond begins on carbon 2, the name of the alkyne containing four carbons is 2-butyne.

Lewis Structure

$$H—\overset{\displaystyle H}{\underset{\displaystyle H}{C}}—C≡C—\overset{\displaystyle H}{\underset{\displaystyle H}{C}}—H$$

Molecular Geometry

$$H\cdots\overset{\displaystyle H}{\underset{\displaystyle H}{C}}—C≡C—\overset{\displaystyle H}{\underset{\displaystyle H}{C}}\cdots H$$

Condensed Structural Formula

CH$_3$CCCH$_3$

Line Drawing

EXERCISE

4·3

For each of the following compounds, draw the Lewis structure.

1. ethyne

2. propyne

3. 2-butyne

4. 1-pentyne

5. 1-hexyne

6. 3-phenyl-butyne

7. 6-methyl-3-octyne

8. 2-hexyne

9. 3,3-dimethyl-4-octyne

10. 4-octyne

EXERCISE
4·4

For each of the compounds in Exercise 4-3, draw the molecular geometry.

1. ethyne

2. propyne

3. 2-butyne

_____

4. 1-pentyne

_____

5. 1-hexyne

_____

6. 3-phenyl-butyne

_____

7. 6-methyl-3-octyne

_____

8. 2-hexyne

_____

9. 3,3-dimethyl-4-octyne

_____

10. 4-octyne

_____

EXERCISE
4·5

For each of the compounds in Exercise 4-3, draw the condensed chemical structure.

1. ethyne

_____

2. propyne

_____

3. 2-butyne

_____

4. 1-pentyne

_____

5. 1-hexyne

_____

6. 3-phenyl-butyne

_____

7. 6-methyl-3-octyne

_____

8. 2-hexyne

_____

9. 3,3-dimethyl-4-octyne

_____

10. 4-octyne

_____

EXERCISE
4·6

For each of the compounds in Exercise 4-3, draw the line drawing.

1. ethyne

2. propyne

3. 2-butyne

4. 1-pentyne

5. 1-hexyne

6. 3-phenyl-butyne

7. 6-methyl-3-octyne

8. 2-hexyne

9. 3,3-dimethyl-4-octyne

10. 4-octyne

Isomers also form with alkynes. Both isomers have the same formula, but their structures are different. Two isomers of C_8H_{14} are shown below.

(i)

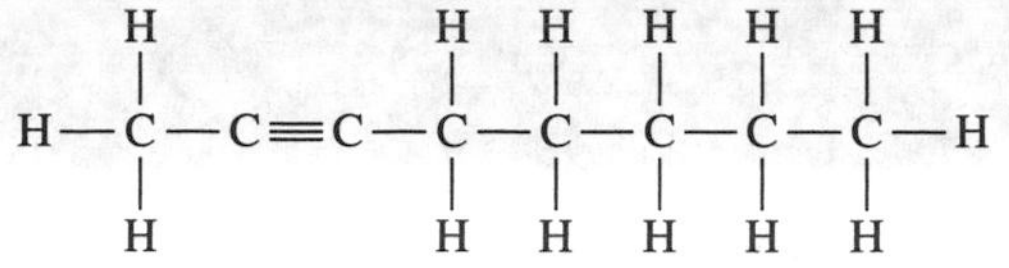

$$H-\underset{\underset{H}{|}}{\overset{\overset{H}{|}}{C}}-C\equiv C-\underset{\underset{H}{|}}{\overset{\overset{H}{|}}{C}}-\underset{\underset{H}{|}}{\overset{\overset{H}{|}}{C}}-\underset{\underset{H}{|}}{\overset{\overset{H}{|}}{C}}-\underset{\underset{H}{|}}{\overset{\overset{H}{|}}{C}}-\underset{\underset{H}{|}}{\overset{\overset{H}{|}}{C}}-H \quad \text{Lewis structure}$$

line drawing

or

(ii)

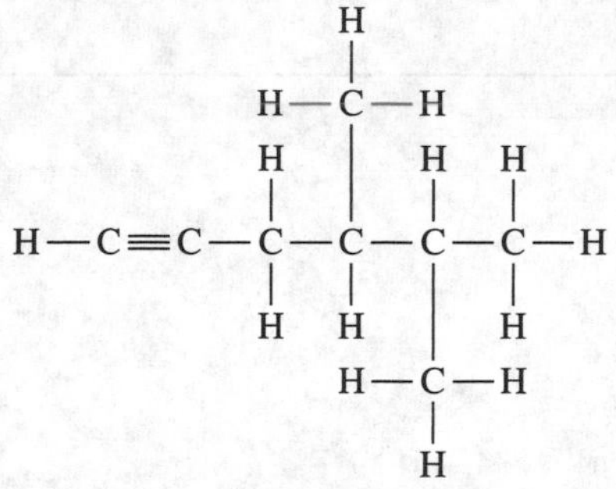

$$H-C\equiv C-\underset{\underset{H}{|}}{\overset{\overset{H-\overset{\overset{H}{|}}{C}-H}{|}}{C}}-\underset{\underset{H}{|}}{\overset{\overset{H}{|}}{C}}-\underset{\underset{H-\overset{\overset{}{|}}{C}-H}{|}}{\overset{\overset{H}{|}}{C}}-\underset{\underset{H}{|}}{\overset{\overset{H}{|}}{C}}-H \quad \text{Lewis structure}$$

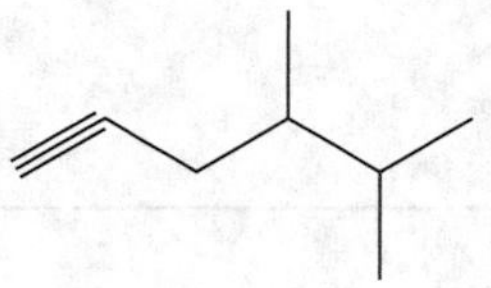

line drawing

Both examples have eight carbon atoms and 14 hydrogen atoms, but their names are different. In the first drawing there are eight carbons in the longest chain, and the triple bond is on the second carbon. The name of this compound is 2-octyne. The second drawing has six carbon atoms in the longest chain, it has a methyl on the fourth and fifth carbons, and the triple bond is on the first carbon. The name of this compound is 4,5-dimethyl-1-hexyne.

EXERCISE
4·7

For each of the following alkynes, draw two possible structural isomers, and then, for each isomer, give the name of the compound, the condensed structural formula, and the line drawing.

1. C_5H_8
   Isomer 1
   Drawing

Name   _____

Structural formula   _____

Line drawing

_____

Isomer 2
Drawing

_____

Name   _____

Structural formula   _____

Line drawing

_____

2. C_7H_{12}

Isomer 1

Drawing

_____

Name

Structural formula

Line drawing

Isomer 2
Drawing

Name

Structural formula

Line drawing

3. $C_{10}H_{18}$

Isomer 1

Drawing

Name _____

Structural formula _____

Line drawing

_____

Isomer 2
Drawing

_____

Name _____

Structural formula _____

Line drawing

_____

4. C_8H_{14}

Isomer 1

Drawing

_____

Name _____

Structural formula _____

Line drawing

_____

Isomer 2
Drawing

_____

Name _____

Structural formula _____

Line drawing

_____

5. C_6H_{10}

Isomer 1

Drawing

_____

Name   _____

Structural formula   _____

Line drawing

_____

Isomer 2
Drawing

_____

Name   _____

Structural formula   _____

Line drawing

_____

*For each of the following line drawings of an alkyne, write the molecular formula, the name
of the compound, and the condensed structural formula.*

1.

Molecular formula   _____

Name   _____

Condensed structural formula   _____

2.

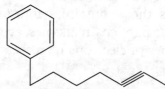

Molecular formula   _____

Name   _____

Condensed structural formula   _____

3.

Molecular formula   _____

Name   _____

Condensed structural formula   _____

4.

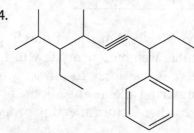

Molecular formula   _____

Name   _____

Condensed structural formula   _____

5.

Molecular formula _____

Name _____

Condensed structural formula _____

Alkyne Properties

Acetylene, the common name for the simplest alkyne, 1-ethyne (C_2H_2), has 14 electrons. Acetylene is a gas used in combination with oxygen gas to produce a high-temperature flame used in welding steel.

$$H—C\equiv C—H$$

One of the main intermolecular forces present in the liquid and solid forms of alkynes are London dispersion forces. For this reason, the lighter-weight alkynes, those containing four or fewer carbons, exist as gases at room temperature. When comparing straight chain alkynes, acetylene, propyne, and 1-butyne, the boiling points increase as the number of electrons increase and the molar mass increases.

Compound	Formula	Boiling point (°C)
Acetylene	HC≡CH	−84
Propyne	HC≡CCH$_3$	−23.2
1-Butyne	HC≡CCH$_2$CH$_3$	8.1

Which compound will have the higher boiling point, 1-butyne (C_4H_{10}) or 1-pentyne? Explain. Because 1-pentyne has more electrons and a greater molecular weight, 1-pentyne will have a higher boiling point compared to 1-butyne. This means there is less force to attract any one molecule to another molecule by dispersion forces. Note, comparing alkene and alkyne properties is not based simply on the number of electrons. There are other factors affecting properties like pi bond interactions which we will not cover here.

$$H—C\equiv C—H$$

Predict which liquid might have the higher boiling point, 2-hexyne or 2-octyne? Explain.

Classification of Alkynes

Alkynes can be classified in terms of the number of groups substituted to the carbons having the triple bond. For alkynes, this is simple: If the carbon that has the triple bond has a hydrogen, the alkyne is said to be monosubstituted.

$$\downarrow$$
$$H-C\equiv C-R$$

In the structure above, R represents an additional "C" chain. The arrow points to a carbon at the beginning of a triple bond.

Another way to classify this compound is as a "terminal alkyne" because the triple bond is at the start of the compound. In the next example, the triple bond is not on a terminal carbon.

$$R-C\equiv C-R$$

If the triple bond has R's on both sides (carbon chains), the triple bond is called an internal alkyne and is disubstituted.

Classify each of the following alkynes as monosubstituted and/or or disubstituted.

1.
$$H-C\equiv C-\overset{\overset{\displaystyle H}{|}}{\underset{\underset{\displaystyle H}{|}}{C}}-\overset{\overset{\displaystyle H}{|}}{\underset{\underset{\displaystyle H}{|}}{C}}-H$$

_____

2.
$$H-\overset{\overset{\displaystyle H}{|}}{\underset{\underset{\displaystyle H}{|}}{C}}-C\equiv C-\overset{\overset{\displaystyle H}{|}}{\underset{\underset{\displaystyle H}{|}}{C}}-H$$

_____

3.
$$H-\overset{\overset{\displaystyle H}{|}}{\underset{\underset{\displaystyle H}{|}}{C}}-\overset{\overset{\displaystyle H}{|}}{\underset{\underset{\displaystyle H}{|}}{C}}-C\equiv C-H$$

_____

4.
```
     H   H        H
     |   |        |
H — C — C — C ≡ C — C — H
     |   |        |
     H   H        H
```

5.
```
                 H   H
                 |   |
H — C ≡ C — C ≡ C — C — C — H
                 |   |
                 H   H
```

Some compounds have both triple bonds and double bonds. In this case, the numbering is done so the multiple bonds are on the lowest numbered carbons, although preference is given to a double bond having the lower number. A hyphen is inserted between sections of the name.

```
        H   H   H   H
        |   |   |   |       H
H — C ≡ C — C — C — C — C = C
        |   |   |           H
        H   H   H
```

or

There are seven carbons in the chain, so part of the name will be hept-. When numbered from left to right, the triple bond is on the first carbon and the double bond is on the sixth carbon. But when numbered right to left, the double bond is on the first carbon and the triple bond is on the sixth carbon. Preference is given to the double bond being on the first carbon, so the name of this compound is 1-hepten-6-yne.

Alkyl groups can also be added. These are treated the same way as with other compounds, appearing in alphabetical order in front of the base names.

```
        H   H   H   H   H       H
        |   |   |   |   |       |
H — C ≡ C — C — C — C — C = C — C — H
        |   |   |   |       |   |
        H   |   H   H       H   H
        H — C — H
            |
            H
```

This compound would be called 6-methyl-2-nonen-8-yne.

For each of the following compounds, make a line drawing, and then draw the Lewis structure.

1. 3,4-heptadiene-6-yne

2. 3-chloro-4,4-dimethyl-1-nonen-6-yne

3. 3-ethyl-5-methyl-1,6,8-decatriyne

Name each of the following compounds.

1.

$$H-\overset{\overset{\displaystyle H}{|}}{\underset{\underset{\displaystyle H}{|}}{C}}-\overset{\overset{\displaystyle H}{|}}{C}=C-C=\overset{\overset{\displaystyle H}{|}}{C}-C\equiv C-\overset{\overset{\displaystyle H}{|}}{\underset{\underset{\displaystyle H}{|}}{C}}-H$$

2.

$$H-\overset{\overset{\displaystyle H}{|}}{\underset{\underset{\displaystyle H}{|}}{C}}-\overset{\overset{\displaystyle H}{|}}{C}=C-C\equiv C-\overset{\overset{\displaystyle H-\overset{\displaystyle H}{|}-H}{|}}{C}-H$$

3. $HC\equiv CC(CH_3)_2CH_2C\equiv CH$

Alcohols

Organic compounds can be composed of more than just carbon and hydrogen. Many of these other compounds can be categorized by the presence a specific group of additional atoms substituting for H atoms. These substituted groups are called functional groups as they impart new specific "functions" or properties to the molecules. Alcohols are one class of organic compounds with a functional group. Alcohols have the general formula ROH, where R represents the carbon chain and -OH is called a hydroxyl group. The simplest alcohol is methanol, CH_3OH, which is also known as methyl alcohol or wood alcohol. Methanol can be viewed as being derived from a methane molecule by substituting a hydroxyl group for one of the hydrogen atoms. Methanol has the following Lewis structure:

methane   methanol

Note that there are two lone electron pairs on the oxygen atom, just as there are in water, H_2O. The bond angle of the C-O-H bond is approximately 108°, larger than the H-O-H bond angle in water, but smaller than the H-C-H bond angle in methane. The importance of this will become clear in the section on properties of alcohols.

Although not all -OH groups in alcohols are located at one end of the molecule, for now, practice changing the parent compound into an alcohol by substituting the parent compound into an alcohol by placing the -OH on the terminal carbon.

Draw the parent molecule, and then substitute an -OH functional group for the terminal hydrogen to produce an alcohol.

PARENT DRAWING   ALCOHOL DRAWING

1. ethane   _____   _____

2. propane   _____   _____

3. butane   _____   _____

4. pentane   _____   _____

5. octane   _____   _____

Nomenclature and Structure of Alcohols

There are two ways of naming alcohols. One way is to name them with common names, and the other way is to use the IUPAC standard, the standard for chemists. You need to recognize both ways because you will see both names—even in the supermarket. Both methods of naming alcohols are described below.

Simple alcohols are commonly named by identifying the parent alkane, naming the alkyl group present, and then adding the word "alcohol." Remember that a methane minus a hydrogen is called a methyl alkyl group, and an ethane minus a hydrogen is called an ethyl alkyl group and so on with longer alkane chains. For example, the following drawing shows how to take the parent methane, identify the methyl group, and then name the alcohol.

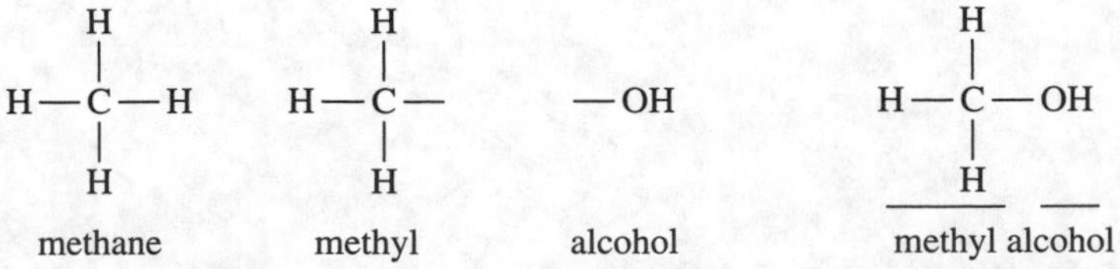

| methane | methyl | alcohol functional group | methyl alcohol or methanol |

The IUPAC name, methanol, is also logically derived from the parent alkane, methane. The terminal "e" in methane is replaced with the suffix -ol.

$$\text{methane} \quad \rightarrow \quad \text{methanol}$$

Another example is ethanol or ethyl alcohol, CH_3CH_2OH, which is known commonly as grain alcohol. Ethanol has the following Lewis structure:

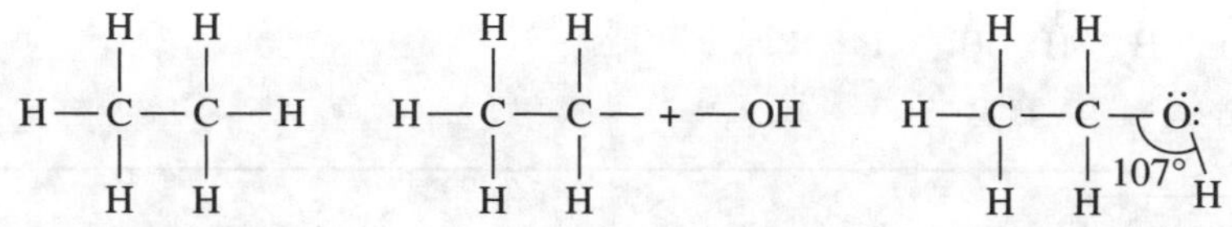

ethane   ethyl   alcohol   ethyl alcohol
                         functional   or
                         group   ethanol

Notice that there are two lone electron pairs on the oxygen atom, and the bond angle of the C-O-H bond is approximately 107°.

$$\text{ethane} \qquad \text{ethanol} \qquad \rightarrow \textbf{ethyl alcohol}$$

The structure of ethyl alcohol or ethanol can also be drawn as follows:

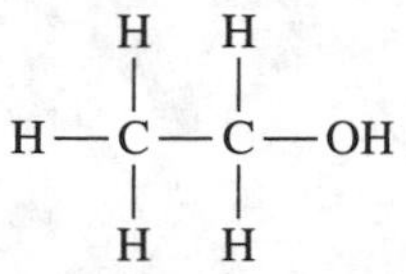

If writing a formula, the -OH should be kept together to help identify the functional group. The ethanol above would be C_2H_5OH instead of C_2H_6O.

EXERCISE
5·2

For each of the following alcohols; write the formula, the common name, and the IUPAC name.

STRUCTURE	FORMULA	COMMON NAME	IUPAC NAME
1. 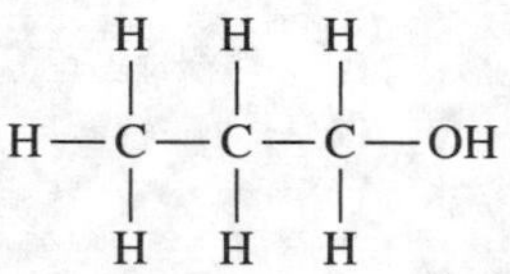 (structure)	_____	_____	_____
2. 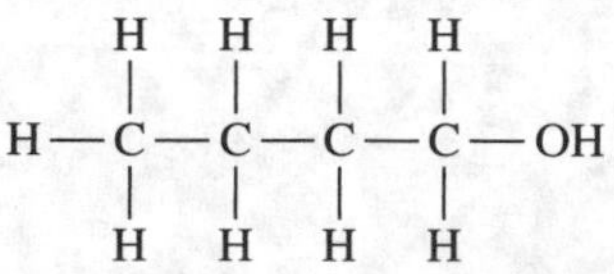 (structure)	_____	_____	_____
3. 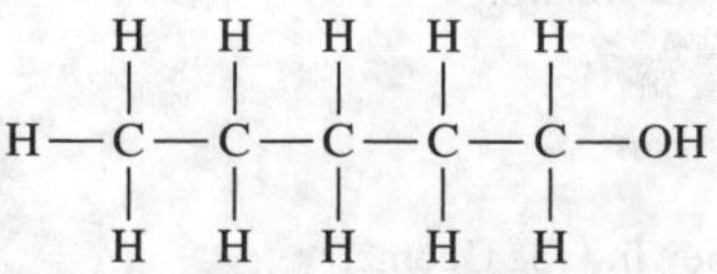 (structure)	_____	_____	_____

4.

$$H-\overset{\overset{\displaystyle H}{|}}{\underset{\underset{\displaystyle H}{|}}{C}}-\overset{\overset{\displaystyle H}{|}}{\underset{\underset{\displaystyle H}{|}}{C}}-\overset{\overset{\displaystyle H}{|}}{\underset{\underset{\displaystyle H}{|}}{C}}-\overset{\overset{\displaystyle H}{|}}{\underset{\underset{\displaystyle H}{|}}{C}}-\overset{\overset{\displaystyle H}{|}}{\underset{\underset{\displaystyle H}{|}}{C}}-\overset{\overset{\displaystyle H}{|}}{\underset{\underset{\displaystyle H}{|}}{C}}-OH$$

_____   _____   _____

5.

$$H-\overset{\overset{\displaystyle H}{|}}{\underset{\underset{\displaystyle H}{|}}{C}}-\overset{\overset{\displaystyle H}{|}}{\underset{\underset{\displaystyle H}{|}}{C}}-\overset{\overset{\displaystyle H}{|}}{\underset{\underset{\displaystyle H}{|}}{C}}-\overset{\overset{\displaystyle H}{|}}{\underset{\underset{\displaystyle H}{|}}{C}}-\overset{\overset{\displaystyle H}{|}}{\underset{\underset{\displaystyle H}{|}}{C}}-\overset{\overset{\displaystyle H}{|}}{\underset{\underset{\displaystyle H}{|}}{C}}-\overset{\overset{\displaystyle H}{|}}{\underset{\underset{\displaystyle H}{|}}{C}}-\overset{\overset{\displaystyle H}{|}}{\underset{\underset{\displaystyle H}{|}}{C}}-OH$$

_____   _____

We can represent the structure of alcohols in several different ways. It is important to be able to recognize and visualize a molecule using a variety of different representations. These are the same representations that are used with alkanes and alkenes.

First review the Lewis structure and molecular geometry for methanol and propanol. The structure and geometry of methanol follow.

Lewis Structure

$$H-\overset{\overset{\displaystyle H}{|}}{\underset{\underset{\displaystyle H}{|}}{C}}-\overset{..}{O}-H$$

methanol
Lewis structure

Molecular Geometry

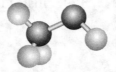

3-D structure showing
Molecular Geometry

Ball and Stick Model showing
the Geometry

Here are these same representations for propanol:

Lewis Structure

$$H-\overset{\overset{\displaystyle H}{|}}{\underset{\underset{\displaystyle H}{|}}{C}}-\overset{\overset{\displaystyle H}{|}}{\underset{\underset{\displaystyle H}{|}}{C}}-\overset{\overset{\displaystyle H}{|}}{\underset{\underset{\displaystyle H}{|}}{C}}-\overset{..}{O}-H$$

Molecular Geometry

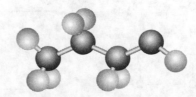

3-D structure showing
Molecular Geometry

Ball and Stick Model showing the Geometry

Draw the Lewis structure and the molecular geometry for each of the following alcohols.

LEWIS STRUCTURE   MOLECULAR GEOMETRY (USE 3-D)

1. ethanol   _____   _____

2. octanol   _____   _____

3. butanol   _____   _____

4. pentanol   _____   _____

5. hexanol   _____   _____

We can also represent the ethanol molecule using a ball-and-stick model:

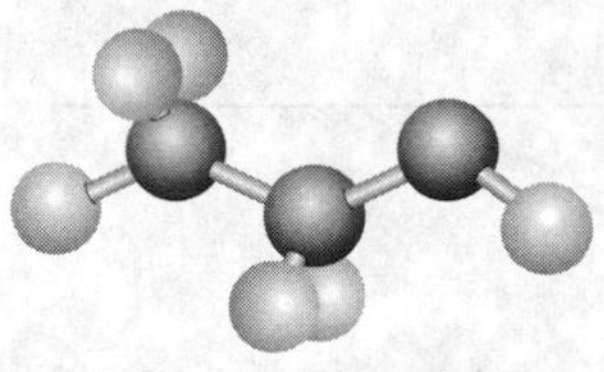

ball-and-stick model

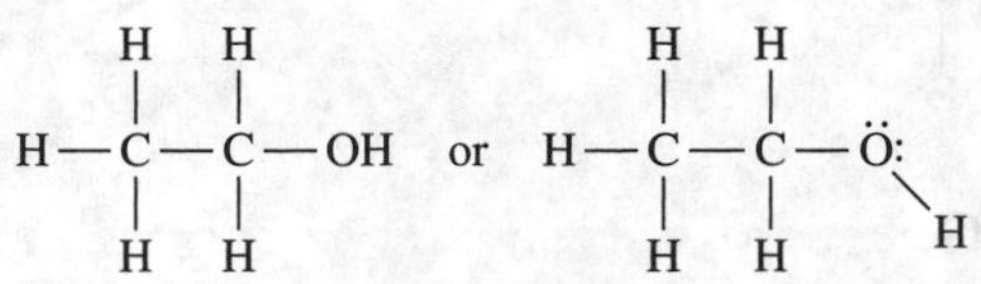

```
    H   H                      H   H
    |   |                      |   |
H — C — C — OH      or    H — C — C — Ö:
    |   |                      |   |      \
    H   H                      H   H       H
```

Lewis structure

```
    H            O — H
    H⟍  |       /
      C — C
    H ⁄        ⟍
    H          H
               H
```

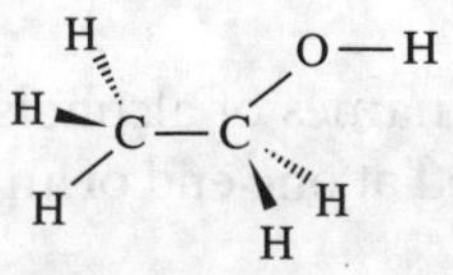

3-D structure

Just as in earlier chapters, the other methods are the condensed formula and the line drawing. Remember that these forms indicate the way the molecule is put together. In the following examples, observe the differences between the condensed formulas and line drawings of ethane and ethanol and the differences between those of ethane and ethanol.

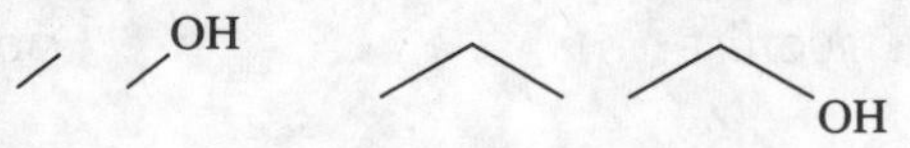

The condensed formulas and line drawings both indicate the carbon in the parent alkane that has the functional group has been substituted for a hydrogen.

EXERCISE
5·4

For each of the following parent alkanes, draw the condensed formula and a line drawing of its alcohol with the functional group on a terminal carbon.

CONDENSED FORMULA   LINE DRAWING

1. propane   _____   _____

2. butane   _____   _____

3. pentane   _____   _____

4. hexane   _____   _____

5. octane   _____   _____

Chemists use the IUPAC names of alcohols. So far, we have looked at simple alcohols with the functional group substituted at one end of an unbranched chain. The functional group can be substituted in other locations.

What are the rules for naming all types of alcohols?

♦ As with alkanes, the parent names of alcohols are based on the longest chain or ring of carbon atoms where one of the carbon atoms is attached to a hydroxyl group. We replace the "e" in the parent alkane and replace it with the suffix -ol.

♦ The position of the hydroxyl group is indicated by the number of the carbon atom to which it is attached. The carbon chain is numbered so that the carbon atom bonded to the hydroxyl group has the lowest number. If more than one hydroxyl group is present, numbers are included for each hydroxyl group.

$$\begin{array}{ccccccc} & H & OH & H & H \\ & | & | & | & | \\ H\!-\!\!&C&\!\!-\!\!&C&\!\!-\!\!&C&\!\!-\!\!C\!-\!H \\ & | & | & | & | \\ & H & H & H & H \end{array}$$

The above chain has four carbons, and the -OH group is on the second carbon, resulting in the name 2-butanol.

$$\begin{array}{ccccccccc} & H & H & H & H & H \\ & | & | & | & | & | \\ H\!-\!\!&C&\!\!-\!\!&C&\!\!-\!\!&C&\!\!-\!\!&C&\!\!-\!\!C\!-\!H \\ & | & | & | & | & | \\ & H & OH & H & OH & H \end{array}$$

Here the longest chain is five carbons, and there are two -OH groups on the second and fourth carbons. Thus di- is used to indicate a total of two -OH groups. The name of this compound is 2,4-dipentanol. Alternatively, the di can be placed in front of the functional group, resulting in 2,4-pentanediol. Note that when the di is placed within the compound, the terminal "e" in pentane is not removed.

If alkyl groups have been added as well as a hydroxyl group, the alkyl groups are added in alphabetical order, and the number of the alcohol is listed last before the parent name. The hydroxyl group has a higher priority than alkyl groups to have the lowest numbered carbon.

$$\begin{array}{ccccccccc} & H & H & H & H & H \\ & | & | & | & | & | \\ H\!-\!\!&C&\!\!-\!\!&C&\!\!-\!\!&C&\!\!-\!\!&C&\!\!-\!\!C\!-\!H \\ & | & | & | & | & | \\ & H & OH & H & & H \\ & & & & | & \\ & & & & H\!-\!C\!-\!H & \\ & & & & | & \\ & & & & H & \end{array}$$

The main chain has five carbons, the second carbon has a hydroxyl group, and the fourth carbon has a methyl group, resulting in the name 4-methyl-2–pentanol.

Changing a Lewis structure to a line drawing is the same only now the functional group will also be shown.

$$\begin{array}{ccccccccc} & H & H & H & H & H \\ & | & | & | & | & | \\ H\!-\!\!&C&\!\!-\!\!&C&\!\!-\!\!&C&\!\!-\!\!&C&\!\!-\!\!C\!-\!H \\ & | & | & | & | & | \\ & H & OH & H & OH & H \end{array}$$

$CH_3CH(OH)CH_2CH(OH)CH_3$

In the example above, 2,4-dipentanol, the condensed formula and the line drawing still show the location of the functional groups.

Write the IUPAC name for each of the following compounds.

1.
```
    H   OH  H   H
    |   |   |   |
H — C — C — C — C — H
    |   |   |   |
    OH  H   H   H
```
_____

2.
```
    H   H   OH  H   H
    |   |   |   |   |
H — C — C — C — C — C — H
    |   |   |   |   |
    H   H   OH  H   H
```
_____

3.
```
    H   OH  H   OH  H   H
    |   |   |   |   |   |
H — C — C — C — C — C — C — H
    |   |   |   |   |   |
    H   H   |   H   H   H
          H — C — H
              |
              H
```
_____

4.
```
    H   H   OH  H   H
    |   |   |   |   |
H — C — C — C — C — C — H
    |   |   |   |   |
    H   |   H   H   H
    H — C — H
        |
    H — C — H
        |
        H
```
_____

5.
```
    H   H   H   H
    |   |   |   |
H — C — C — C — C — H
    |   |   |   |
    OH  H   |   H
          H — C — H
              |
              H
```
_____

What if the structure is not a chain but a ring? The third rule for naming alcohols indicates how to name these type of alcohols. Again the parent name of the alkane (or alkene or alkyne) is retained as the base name.

When the hydroxyl group is attached to a ring, the ring is numbered starting with the atom bonded to the hydroxyl group. The parent ring is named a cyclo___ol. The number "1" is not used to indicate the location of the hydroxyl group because that would be redundant.

```
              OH
     H        |        H
      \       C       /
   H — C           C — H
      /   \       /   \
     H     C — H       H
          / \
     H — C   C — H
          \ /   \
           C     H
          / \
         H   H
```

In the above ring, there are six carbons and one -OH group. This would be named cyclo-hexanol. These compounds are similar to other compounds wherein more than one type of group have been added to the ring.

For example, the following molecule is named 3,4-dimethylcyclohexanol.

There are six carbons in the ring, so the base is cyclohexanol. Counting with the carbon that has the hydroxyl group as carbon 1, the two methyl groups are on carbons 3 and 4, resulting in the name 3,4-dimethylcylohexanol. If the ring is a benzene (C_6H_6) with an -OH in place of one of the H's, the compound has a special name and is called a phenol (not benzyl alcohol).

EXERCISE
5·6

Write the correct IUPAC name for each of the following compounds.

1.

_____

2.

_____

3.

_____

4. 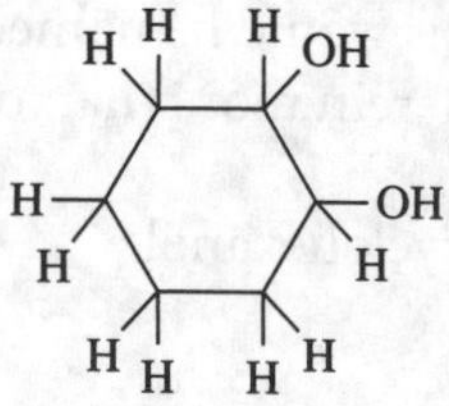

5. 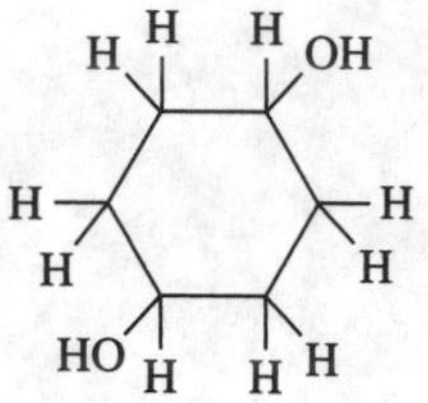

EXERCISE

5·7

Write the condensed formula and line drawing for each of the compounds in Exercise 5-6.

	CONDENSED FORMULA	LINE DRAWING
1. 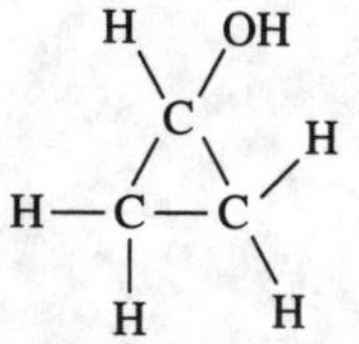		
2. 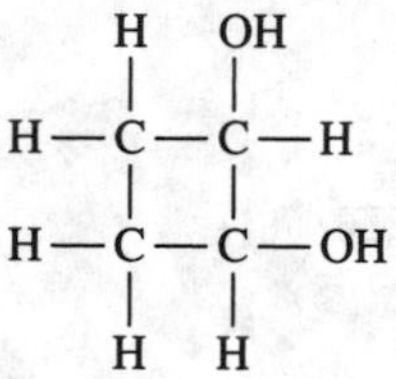		
3. 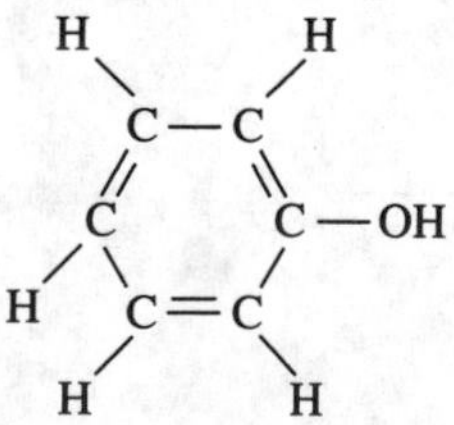		
4. 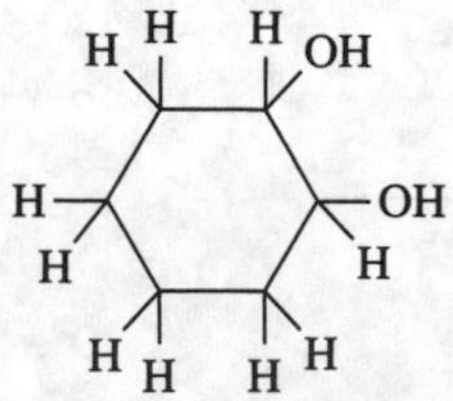		

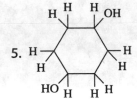

5.

EXERCISE
5·8

Draw the Lewis structure for each of the following compounds.

1. cyclobutanol

2. 3-ethyl-cyclohexanol

3. 2-methyl-cyclopropanol

4. 2,4-dimethyl-3-pentanol

_____

5. 1,3-diphenol

_____

Classification of Alcohols

Alcohols are also classified into three groups—primary alcohols (1°), secondary alcohols (2°), and tertiary alcohols (3°)—by the number of alkyl groups attached to the carbon atom bonded to the hydroxyl (-OH) group.

$$
\begin{array}{ccc}
\text{H} & \text{R} & \text{R} \\
| & | & | \\
\text{R—C—OH} & \text{R—C—OH} & \text{R—C—OH} \\
| & | & | \\
\text{H} & \text{H} & \text{R} \\
\text{primary} & \text{secondary} & \text{tertiary} \\
\text{alcohol} & \text{alcohol} & \text{alcohol}
\end{array}
$$

In classifying alcohols, the names of the compounds are modified. Both the common and IUPAC names are useable names for the same compound. This is one reason why organic naming can be confusing.

If the carbon atom attached to the hydroxyl group has only one other carbon atom attached to it, the carbon is called a primary carbon, and the alcohol is classified as a primary alcohol.

$$
\begin{array}{cc}
\text{H H H} & \text{H H H H} \\
|~~|~~| & |~~|~~|~~| \\
\text{H—C—C—C—OH} & \text{H—C—C—C—C—OH} \\
|~~|~~| & |~~|~~|~~| \\
\text{H H H} & \text{H H H H} \\
n\text{-propanol} & n\text{-butanol}
\end{array}
$$

n-Propanol is a 1° alcohol. *n*-Butanol is a 1° alcohol.

> The *n*- is shorthand for "normal" and indicates a straight chain with the alcohol group(s) at either or both ends. The prefix *sec*- (or *s*-) is used for secondary alcohols, and *tert*- (or *t*-) is used for tertiary alcohols.

If the carbon atom bonded to the hydroxyl group has two other carbon atoms bonded to it, the carbon is called a secondary carbon, and the alcohol is classified as a secondary alcohol. Note the location of the -OH group in the diagram that follows. It is bonded to a carbon that is bonded to two other carbons, so this is a secondary alcohol.

sec-Butyl alcohol is a 2° alcohol.

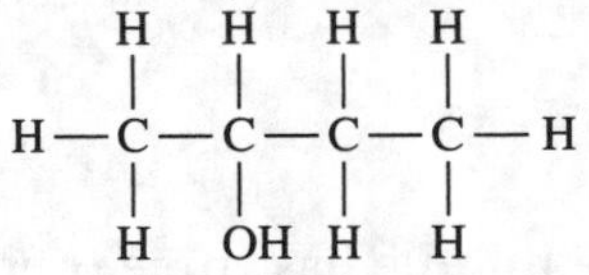

sec-butyl alcohol or 2-butanol
four-carbon chain
Alcohol on carbon attached to two other carbons

sec-Propyl alcohol is a 2° alcohol.

Carbon with-OH functional group is attached to two other carbons.

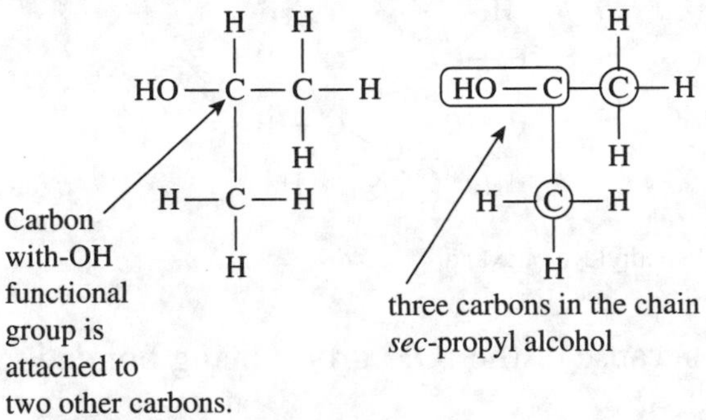

three carbons in the chain
sec-propyl alcohol

Alcohols in the isopropyl group have two methyl groups bonded to the carbon, and the carbon has a single terminal group bonded to it, making their structure unique. An alternative name is frequent for this compound, as it is an isomer of propyl alcohol. Since there is only one possible alternative, sec-propyl alcohol is sold under the name isopropanol or iso propyl alcohol. The following diagram represents the isopropyl group.

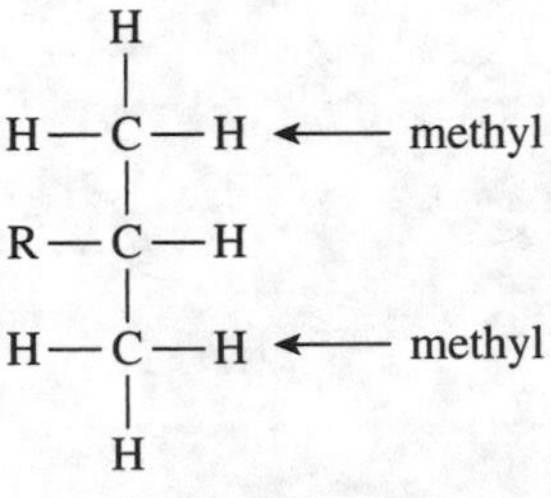

Since the group being added is an -OH group, the following structure represents a molecule of isopropyl alcohol.

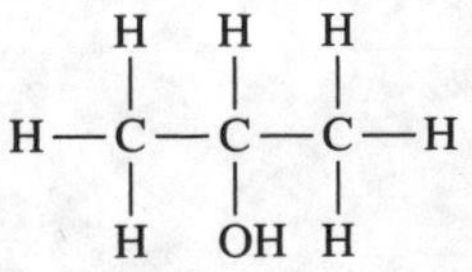

Isopropyl alcohol is a secondary alcohol (It is also called 2-propranol).

Isopropyl alcohol is used as a disinfectant/antiseptic in cleaners for glasses and in medical wipes.

If the carbon atom bonded to the hydroxyl group has three carbon atoms attached to it, then the carbon is called a tertiary carbon, and the alcohol is classified as a tertiary alcohol. This can only happen in alcohols with at least four carbon atoms.

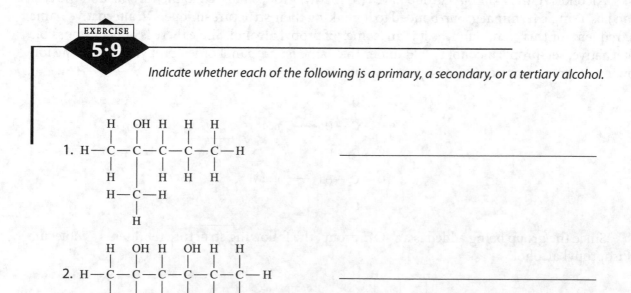

tert-butyl alcohol is also called 2-methyl-2-propanol.

t-Butyl alcohol is a 3° alcohol. Note that the carbon in the middle has three other carbons bonded to it as well as an alcohol (-OH) group.

1-methylcyclohexanol

1-Methylcyclohexanol is a 3° alcohol because of the -OH group being bonded to a carbon that is bonded to three other carbons.

EXERCISE
5·9

Indicate whether each of the following is a primary, a secondary, or a tertiary alcohol.

1.

2.

3. 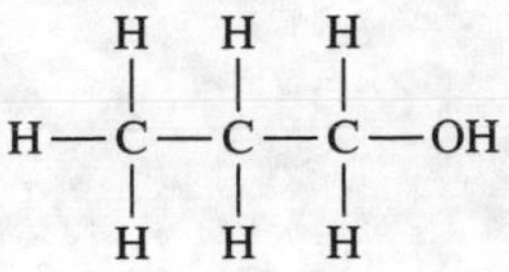
```
     H   H   H
     |   |   |
H —  C — C — C — OH
     |   |   |
     H   H   H
```
_____

4. 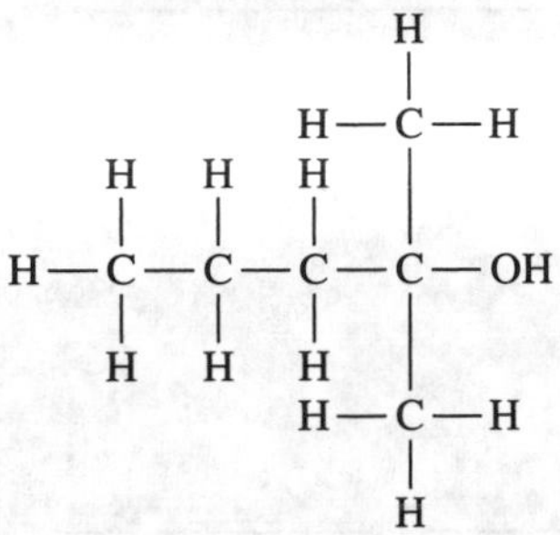
```
                H
                |
            H — C — H
                |
     H   H   H  |
     |   |   |  |
H —  C — C — C — C — OH
     |   |   |
     H   H   H
                |
            H — C — H
                |
                H
```
_____

5.

```
HO    H           HO   H
  \  /              \  /
   C                 C
  / \               / \       OH
     OH         H   C   C
                 \ /   / \
              H — C       H
                 /        |
                H         OH
                          |
              H — C   C — H
                 / \ / \
                H   C   H
                    |
                    H  H
```
_____

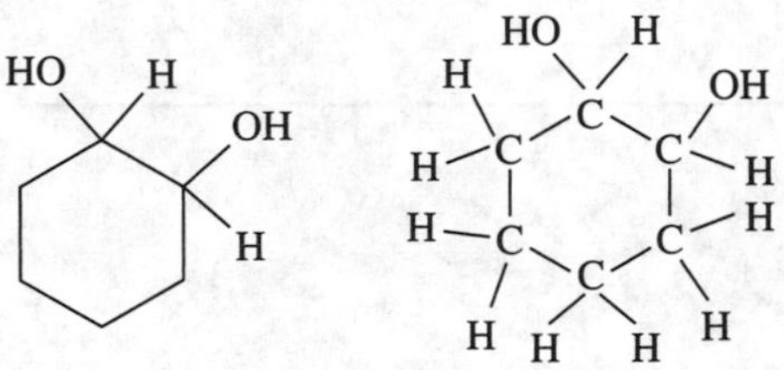

EXERCISE
5·10

Name the following structures.

1. 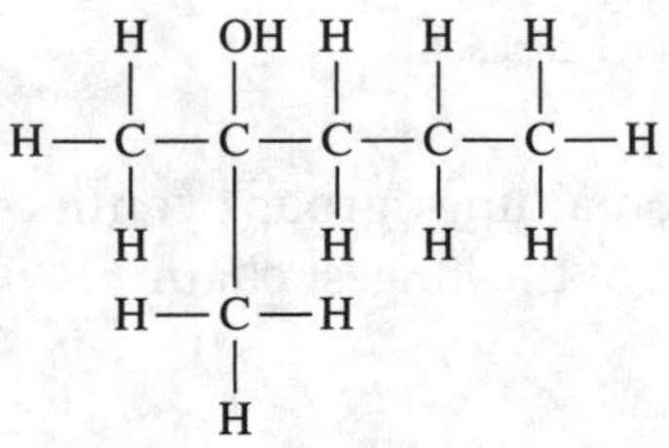
```
     H   OH  H   H   H
     |   |   |   |   |
H —  C — C — C — C — C — H
     |   |   |   |   |
     H   |   H   H   H
         |
     H — C — H
         |
         H
```
_____

2. 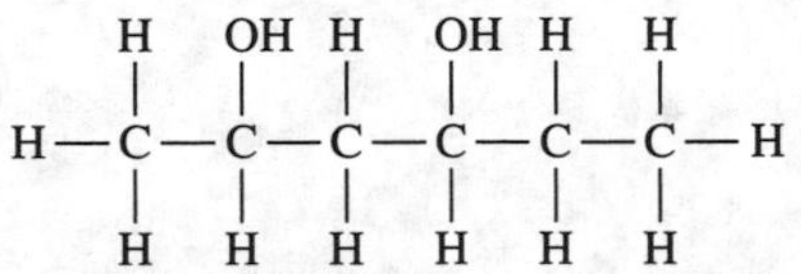
```
     H   OH  H   OH  H   H
     |   |   |   |   |   |
H —  C — C — C — C — C — C — H
     |   |   |   |   |   |
     H   H   H   H   H   H
```
_____

3. 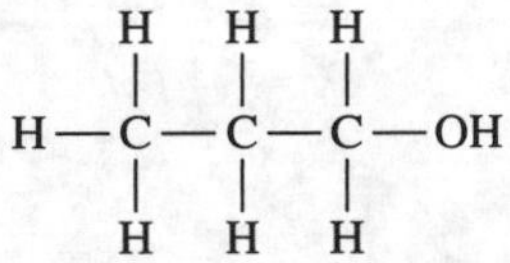
```
     H   H   H
     |   |   |
H —  C — C — C — OH
     |   |   |
     H   H   H
```
_____

4. 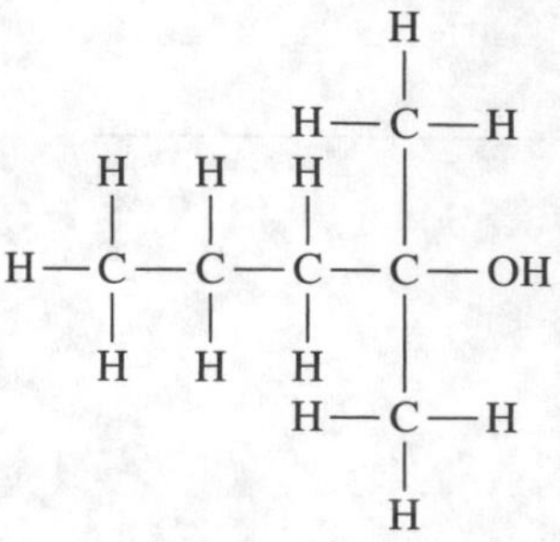 H—C—C—C—C—OH

5.

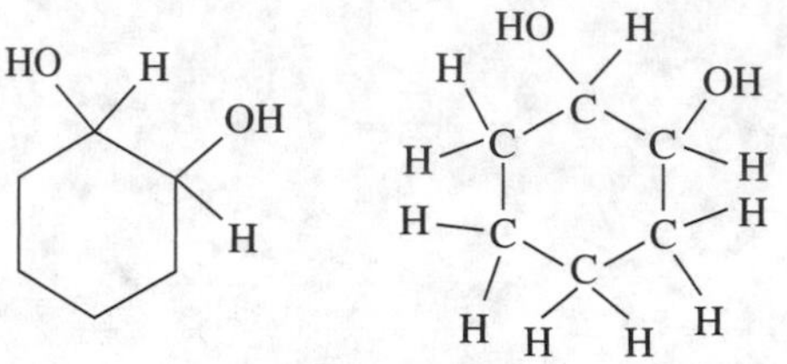

Isomers

Compounds that have the same formula but different names are called isomers. For example, butanol and methylpropanol both have the formula C_4H_7OH, but their structures differ:

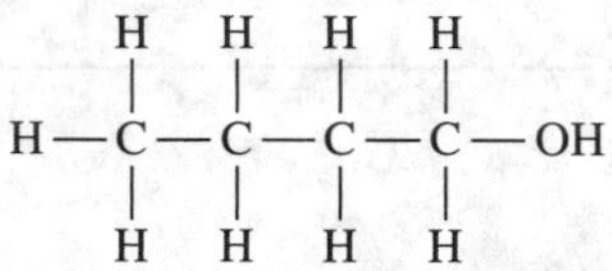

butanol

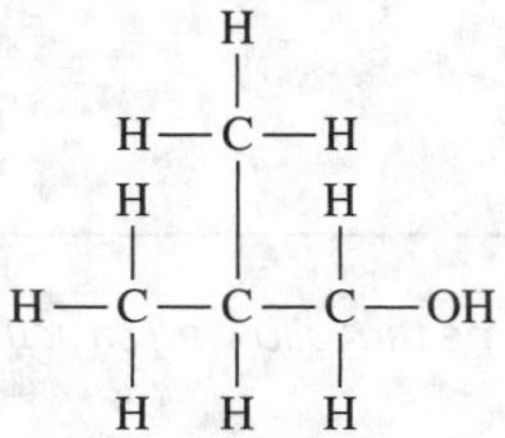

2-methylpropanol

When drawing an isomer, a substitution of a carbon containing "group" cannot be on a terminal carbon because chains can bend and it would be part of the longest chain.

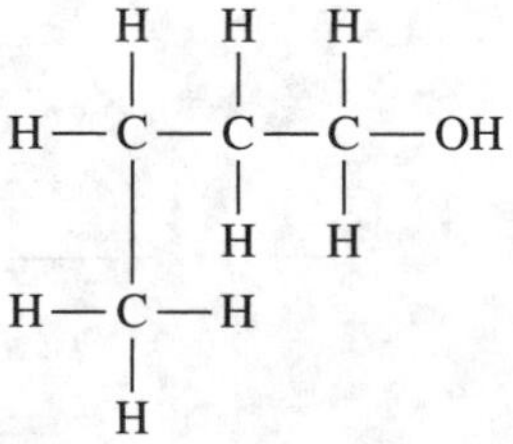

butanol

The above structure is still butanol and is not considered an isomer. This means that to form isomers, the additions must be on internal carbons. Structures showing the addition above or below on the same carbon also do not represent isomers:

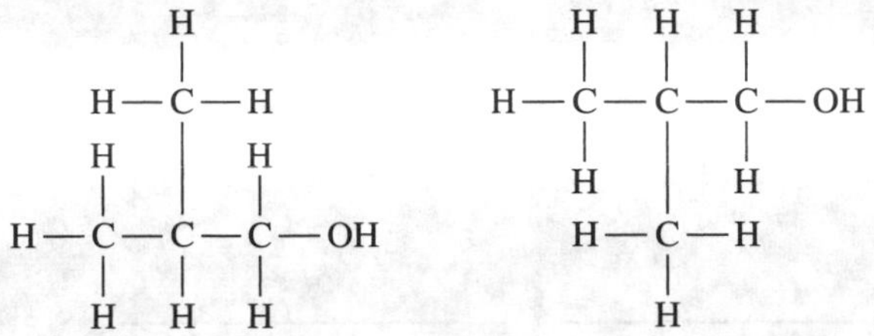

$$
\begin{array}{cc}
\begin{array}{c}
\quad\ \ \ H \\
\quad\ \ \ | \\
H{-}C{-}H \\
\ \ H\ |\ \ H \\
\ \ |\ \ |\ \ | \\
H{-}C{-}C{-}C{-}OH \\
\ \ |\ \ |\ \ | \\
\ \ H\ H\ H
\end{array}
&
\begin{array}{c}
\ \ H\ \ H\ \ H \\
\ \ |\ \ |\ \ | \\
H{-}C{-}C{-}C{-}OH \\
\ \ |\ \ |\ \ | \\
\ \ H\ \ |\ \ H \\
\quad H{-}C{-}H \\
\quad\ \ \ | \\
\quad\ \ \ H
\end{array}
\end{array}
$$

These are both 2-methylpropanol.

5·11

For each of the following alcohols, write the formula of the given alcohol, draw two isomers, and name those isomers.

FORMULA	DRAWINGS	NAME

1. pentanol

_____   _____   _____

2. hexanol

_____   _____   _____

3. heptanol

_____ _____ _____

4. octanol

_____ _____ _____

5. decanol

_____ _____ _____

Alcohol Properties

methanol   ethanol

Note that there are two lone electron pairs on the oxygen atom in both structures and that the bond angle of the C-O-H bond is less than 109°. This is important because it results in alcohols being polar, and having, partly, a similar structure to water.

H—Ö: ◄——— area of partial negative charge δ⁻
    |
    H

area of partial positive charge δ⁺

This characteristic enables alcohols to be dissolved with complete mixing in water, which is also polar.

hydrogen bond
between ethanol
and water

Besides being able to dissolve in water, alcohols form hydrogen bonds with other alcohol molecules and also with water molecules. This hydrogen bonding, being stronger intermolecular forces than the London Dispersion forces which are also present, results in pure alcohols having higher melting points and boiling points than their parent alkane as a consequence of the greater intermolecular force between the molecules.

propane (parent)
dispersion forces only

propanol
dispersion forces
dipole forces
hydrogen bonding

The parent alkanes have dispersion forces based mostly on the number of electrons dispersed in the molecule. Generally, dispersion forces between molecules are weaker than hydrogen bonds between molecules.

hydrogen bonding between
two methanol molecules

When two different alcohols are compared, the one with the greatest intermolecular forces will be found to have the higher boiling point. In general, when comparing two different alcohols, count the number of -OH groups. The alcohol with the greater number of hydroxyl groups has the greater strength of intermolecular forces—and therefore the highest boiling point.

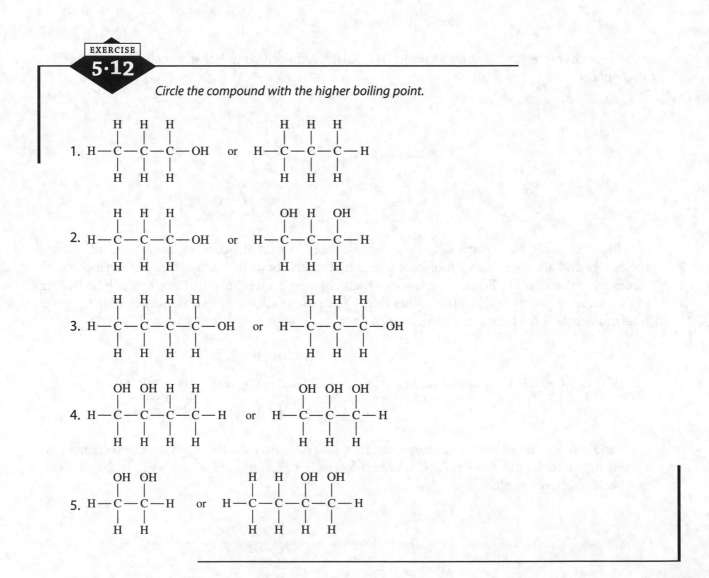

Ethanol has one -OH group; 1,2-ethanediol (also called ethylene glycol) has two -OH groups; and 1,2,3-propanetriol (also called glycerin) has three -OH groups. Glycerin has the highest boiling point of the three compounds.

EXERCISE
5·12

Circle the compound with the higher boiling point.

Not all important types of alcohols are mentioned in this chapter. Some, such as cholesterol (shown below), are very complex and are recognized usually by their common name not their IUPAC name.

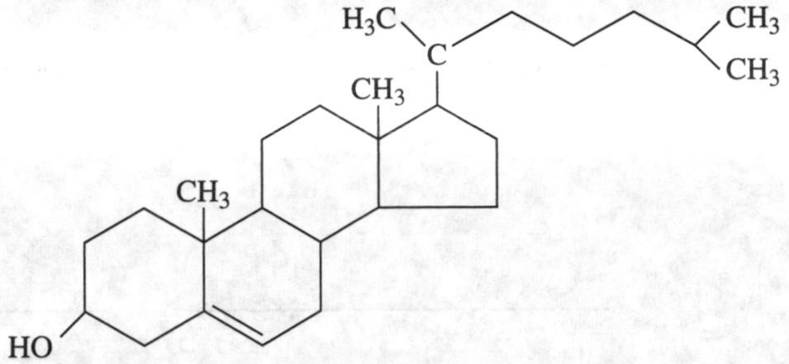

Additional Practice

EXERCISE
5·13

Given the following structures, name each compound and write its formula.

NAME   FORMULA

1.

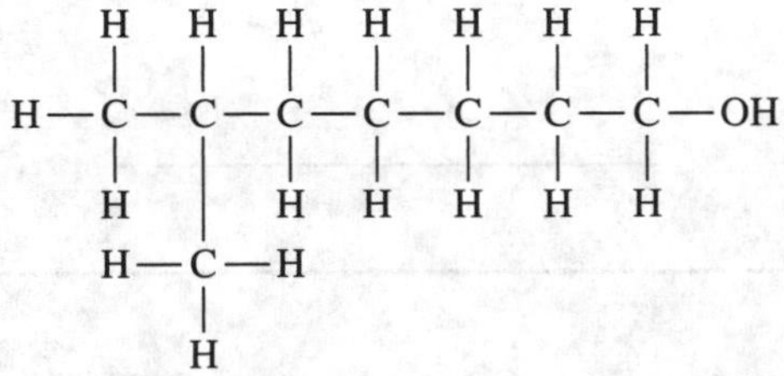

2.

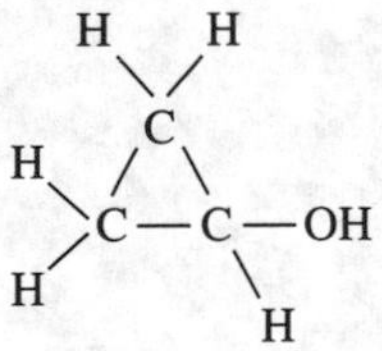

3.

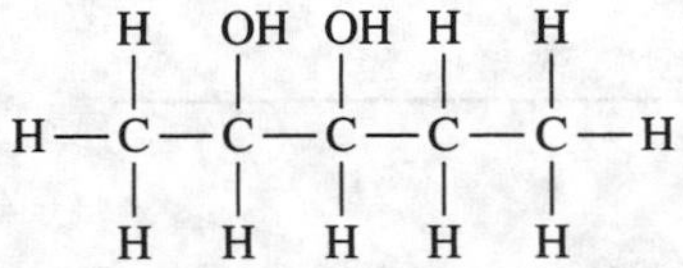

4.

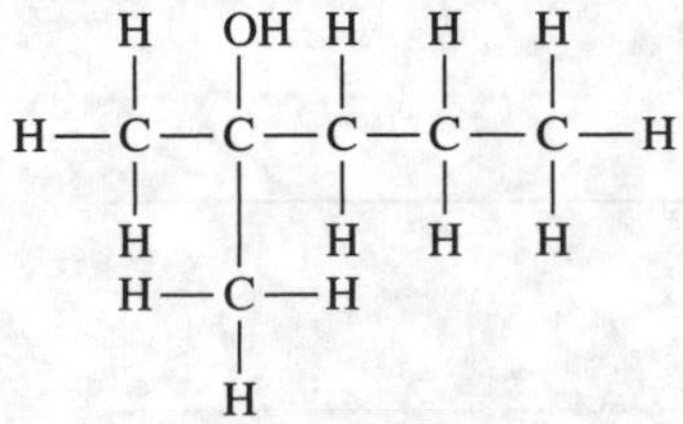

EXERCISE
5·14

For each of the following compounds, draw the structure and write the formula of the compound.

	FORMULA	STRUCTURE
1. methanol	______	______
2. 2,3-pentanediol	______	______
3. 2,2,3-hexanetriol	______	______
4. isopropyl alcohol	______	______

EXERCISE
5·15

Classify the following alcohols as being 1°, 2°, or 3°.

1. 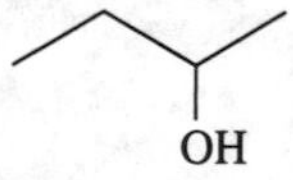 ______

2. 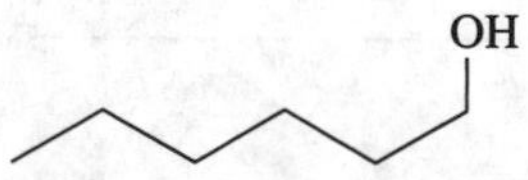 ______

3. 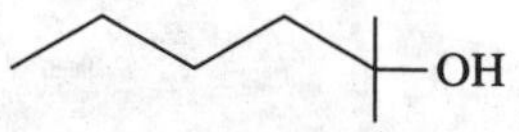 ______

4. 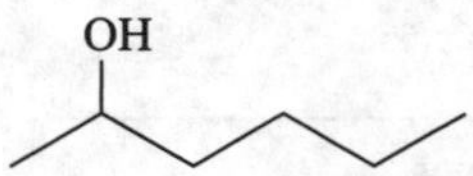 ______

EXERCISE
5·16

For each of the following compounds; draw the Lewis structure.

1. 1,2-ethanediol

2. 1,2,3-propanetriol

3. *n*-octanol

4. 3-phenyl-2-butanol

For each of the following compounds, draw the molecular geometry.

1. methanol

2. propanol

EXERCISE
5·18

For each of the following compounds, draw the condensed chemical structure.

1. 2-methyl-2-pentanol

2. 1,4-pentanediol

3. 2-octanol

4. 1,2,3-propanetriol

EXERCISE

5·19

For each of the following compounds, draw the line drawing.

1. 2-methyl-2-pentanol

2. 1,4-pentanediol

3. 2-octanol

4. 1,2,3-propanetriol

Given the molecular formula of an alcohol, draw two isomers, and for each isomer, write the name of each compound, the condensed structural formula, and the line drawing.

1. C₅H₈ OH

Isomer 1

Drawing

_____

Name_____

Structural formula _____

Line drawing

_____

Isomer 2

Drawing

_____

Name _____

Structural formula _____

Line drawing

_____

2. C_7H_{14} OH

Isomer 1

Drawing

_____

Name _____

Structural formula _____

Line drawing

_____

Isomer 2

Drawing

_____

Name _____

Structural formula _____

Line drawing

---

3. $C_6H_{12}OH$

Isomer 1

    Drawing

Name _____

Structural formula _____

Line drawing

Isomer 2

    Drawing

Name_____

Structural formula_____

Line drawing

_____

4. C_4H_8 OH

Isomer 1

    Drawing

_____

Name_____

Structural formula_____

Line drawing

_____

Isomer 2

    Drawing

_____

Name_____

Structural formula _____

Line drawing

EXERCISE
5·21

Given the line drawing of an alcohol, write the molecular formula, the name of the compound, and the condensed structural formula.

1.

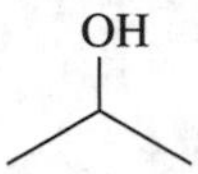

_____

_____

2.

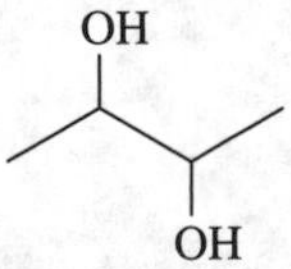

_____

_____

3.

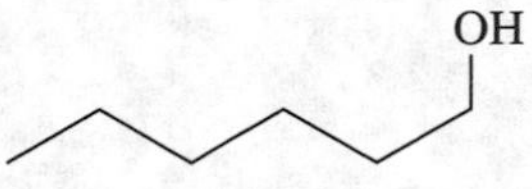

_____

_____

4.

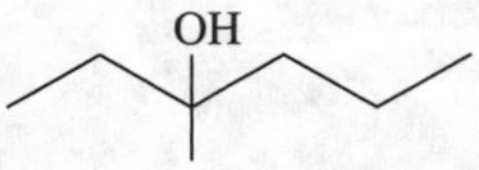

_____

_____

EXERCISE
5·22

Predict which liquid has the higher boiling point, and explain why.

1. *n*-propanol or methanol _____

   _____

2. propane or *n*-propanol _____

3. 1,2,3-propanetriol or 1,2-ethanediol _____

Aldehydes

6

Aldehydes make up a class of organic compounds that have the general structural formula

$$
\begin{array}{c}
:\!O\!: \\
\| \\
R\!-\!C\!-\!H
\end{array}
$$

The main structural feature of an aldehyde is a carbonyl group which consists of a carbon atom double-bonded to an oxygen atom. There are several different organic structures having one or more carbonyl groups. Aldehydes are distinguished as they also have a hydrogen atom bonded to the carbon atom of the carbonyl group. The O-C-H group is called a formyl group and is always on a terminal end of a carbon chain as the carbon atom involved can only form one more pair bond.

$$
\begin{array}{c}
H \quad\;\; :\!O\!: \\
| \qquad \| \\
H\!-\!C\!-\!C\!-\!H \\
| \\
H
\end{array}
$$

ethanal

2 carbons (eth-) single bonds
between carbons (-an-) and
a formal group (-al) equals ethanal.

The Lewis structure for the carbonyl group indicates that there is a double bond between the carbon atom and the oxygen atom and that there are two lone electron pairs on the oxygen atom. To explain the structural shape, we can say that both the carbon atom and the oxygen atom have valence electrons that are sp^2 hybridized. The double bond consists of a sigma bond and a pi bond, just as in between two C atoms in alkenes.

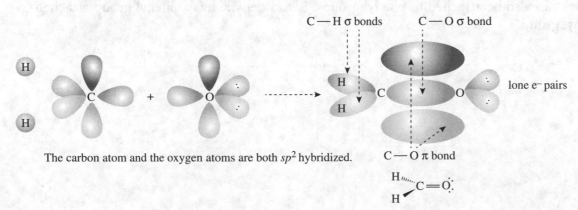

The carbon atom and the oxygen atoms are both sp^2 hybridized.

127

The carbonyl group is a special group characterized by the relatively high electronegativity of the oxygen atom and the relatively low electronegativity of the carbon atom. The oxygen atom draws the electron density in the bond toward itself, creating a partial negative charge on the oxygen and a partial positive charge on the carbon. This results in the carbonyl group being polar. This has great significance for potential chemical reactions since nucleophiles (electron-rich areas with a negative charge that donate an electron pair to electrophiles) seek out the carbon atom, while electrophiles (electron-poor areas with a positive charge that accept electrons from nucleophiles) seek out the oxygen atom. This makes the H atom on the same carbon rather weakly bonded, and consequently, aldehydes are highly reactive.

$$\begin{array}{c} \overset{\delta-}{:O:} \longleftarrow \text{electron-rich area} \\ \| \\ R - C - H \\ \overset{\delta+}{} \longleftarrow \text{electron-poor area} \end{array}$$

The simplest aldehyde has one carbon. It is officially called methanal or commonly formaldehyde. Notice that the official IUPAC naming system is standard and based on the parent alkane just as it was for alcohols.

$$\begin{array}{c} O \\ \| \\ H - C - H \end{array}$$

Ethanal, or more commonly, acetaldehyde, has the Lewis structure

$$\begin{array}{c} H \quad O \\ | \quad \| \\ H - C - C - H \\ | \\ H \end{array}$$

An identifying characteristic to remember is that the aldehyde group must involve a terminal carbonyl group, C=O , with a hydrogen attached to the carbon, C-H.

Aldehydes with several carbon atoms, in general, have pleasant odors and can be found in many natural products. For example, benzaldehyde is the primary component of almond oil. Aldehydes are also responsible for the smells of cinnamon, cherry, and vanilla and are therefore used in many perfumes. Not all aldehyde odors are pleasant, though. Formaldehyde, which has a nasty odor, was once used to preserve biological and anatomical specimens such as frogs. It is known to be quite toxic and is released from some chemical foams used for home insulation. Aldehydes can also be attached to phenyl groups. Benzaldehyde has a phenyl group attached to the formyl group.

As for all organic molecules, there is a systematic way to name the aldehydes. Each aldehyde has a unique IUPAC name. Just as in the naming of alkanes, alkenes, and alcohols, the number of carbons in the longest chain is designated by a prefix. The suffix -al is used to indicate the presence of a formyl group at one end of the carbon chain. The carbon atom in the formyl group is understood to be numbered as the first carbon; so the number "1" is not used in the name. As is true of the alcohols, the location(s) of any substituents are listed with the lowest possible carbon number in the chain.

Example: Name the following organic structure.

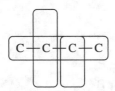

The main chain consists of four carbons with single bonds
and a carbonyl functional group on the end (butanal).

4   3   2   1   Numbering starts from the functional group.

Three methyl groups (— CH$_3$) are added, for the
prefix tri-. They are located on carbons 2, 3, and 3.

2,3,3-trimethyl butanal

The main chain has four carbons with a formyl group on one end (butanal), and there are three methyl groups: one on carbon 2 and two on carbon 3 (2,3,3-trimethyl), resulting in the name 2,3,3-trimethylbutanal.

When the compound name ends in -enal, the "en" indicates that a double bond is present in the compound in addition to a formyl group. An example is 2-propenal. When a double bond and a formyl group are both present, the formal group dictates the numbering of the carbons in the chain.

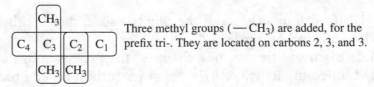

Three carbons (prop-) with double bond
(-en-) on the second bond, and a carbonyl
on the end (-al) yield 2-propenal.

Remember that when double bonds exist, the chain can continue on the same side (*cis-*) or across to the opposite side (*trans-*).

trans-2-butenal   *cis*-2-butenal

A compound ending in –dial indicates that there is an aldehyde group on each end of the carbon chain. The "di" indicates two, and the "al" indicates aldehyde groups.

butandial

When the aldehyde group is attached to a ring, the suffix -carbaldehyde is used in the name.

or

The aldehyde group is attached to a ring, and the ring has six carbons with single bonds, resulting in the name cyclohexanecarbaldehyde.

In all the compounds discussed in previous chapters, the groups substituted in compounds have been mostly limited to methyl-, ethyl- and other carbon-based groups. As mentioned in an earlier chapter, besides the -OH groups of alcohols, atoms of other elements can also be substituted to all those compounds. In naming these substitutions, alphabetical order is used, and an "o" is added to the atom base. Here are some examples (the base name is underlined).

Cl	chlorine	→	chloro-
Br	bromine	→	bromo-
F	fluorine	→	fluoro-
I	iodine	→	iodo-

When writing formulas, as was with alcohols, keeping the functional group together helps to identify the compound. For example, cyclohexanecarbaldehyde would be $C_6H_{10}OCH$.

EXERCISE 6·1

Given the following structures, name each compound and write its formula.

NAME   FORMULA

1.
```
    H   H   H   H   H   O
    |   |   |   |   |   ‖
H — C — C — C — C — C — C — H
    |   |   |   |   |
    H   H   H   H   H
```

_____   _____

2.
```
    H   H   H   H   H   O
    |   |   |   |   |   ‖
H — C — C — C — C — C — C — H
    |   |   |   |   |
    H   H   Br  H   Br
```

_____   _____

3.
```
    H   H   H   H   H   O
    |   |   |   |   |   ‖
H — C — C — C — C — C — C — H
    |   |   |   |   |
    H   Br  H   |   |
        |       |   |
    H — C — H   H — C — H
        |           |
        H           H
```

_____   _____

4.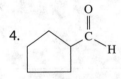

_____   _____

5.
```
    H   H   H   O
    |   |   ‖   ‖
H — C — C = C — C — H
    |
    H
```

_____   _____

EXERCISE 6·2

For each of the following compounds, draw the structure and write the formula of the compound.

STRUCTURE   FORMULA

1. 3-chloropentanal

_____   _____

2. 2-ethylbutanal

_____   _____

3. propanal

_____   _____

4. cyclohexanecarbaldehyde

_____   _____

5. 2,2-diobromo-4-methylhexanal

_____   _____

We can represent the structure of aldehydes in several different ways, just as was done for all other functional group compounds studied so far. It is important to be able to recognize and visualize a molecule using a variety of different representations. These are the same representations that are used with alkanes and alkenes. The difference is the presence of the formyl group.

For butanal, these representations include

Lewis Structure

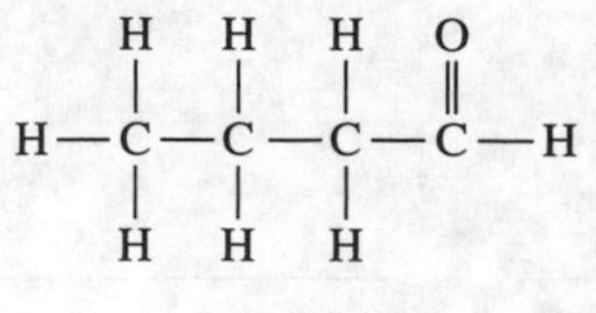

$$H-\overset{\overset{\displaystyle H}{|}}{\underset{\underset{\displaystyle H}{|}}{C}}-\overset{\overset{\displaystyle H}{|}}{\underset{\underset{\displaystyle H}{|}}{C}}-\overset{\overset{\displaystyle H}{|}}{\underset{\underset{\displaystyle H}{|}}{C}}-\overset{\overset{\displaystyle O}{||}}{C}-H$$

Molecular Geometry

Condensed Structural Formula

$CH_3CH_2CH_2CHO$ or $CH_3CH_2CH_2CH$
$\overset{\displaystyle O}{\overset{\|}{}}$

Line Drawing

Remember that a "line drawing" shows just the minimum amount of information. If the molecule is based on a hydrocarbon, the carbon connections are shown, and any other atoms, but not the hydrogens. The only exception is if the functional group has a hydrogen.

EXERCISE
6·3

For each of the following compounds, draw the Lewis structure.

1. 2-methylcyclopentanecarbaldehyde

2. *cis*-2-pentenal

3. 2-4-dichloro-2-methylhexanal

4. 5-methyl-3-hexenal

5. *n*-octanal

EXERCISE

6·4

For each of the following compounds, draw the molecular geometry.

1. 2-ethyl-4-methylpentanal

2. 3,3-dibromo-4-methylhexanal

_____

3. 3-phenylpropanal

_____

4. *n*-pentanedial

_____

5. 4-heptenal

_____

EXERCISE

6·5

For each of the following compounds, draw the condensed chemical structure.

1. 2-ethyl-4-methylpentanal

____

2. 3,3-dibromo-4-methylhexanal

____

3. 3-phenylpropanal

____

4. *n*-pentanedial

____

5. 4-heptenal

____

EXERCISE

6·6

For each of the following compounds, draw the line drawing.

1. 2-ethyl-4-methylpentanal

2. 3,3-dibromo-4-methylhexanal

3. 3-phenylpropanal

4. *n*-pentanedial

5. 4-heptenal

EXERCISE

6·7

For each of the following aldehydes, draw the structure for one isomer, write the name of the compound, and draw its line drawing.

	STRUCTURE	NAME	LINE DRAWING
1. C_2H_4BrCHO	______	______	______
2. $C_5H_{11}CHO$	______	______	______
3. $C_8H_{17}CHO$	______	______	______
4. $C_8H_{15}Cl_2CHO$	______	______	______

For each of the following aldehydes, write the molecular formula, the name of the compound, and the condensed structural formula.

	MOLECULAR FORMULA	NAME	CONDENSED STRUCTURE

1.

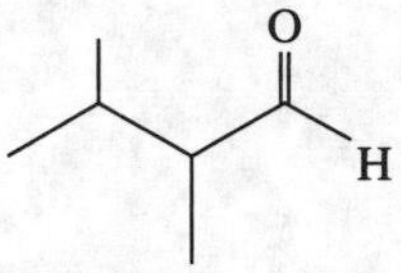

_____ _____ _____

2.

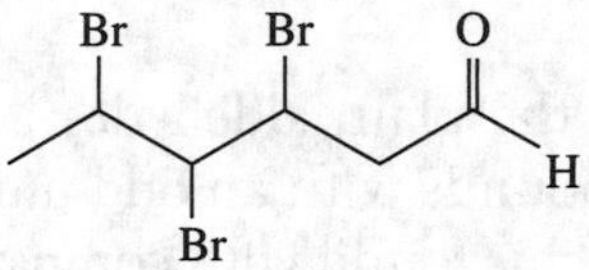

_____ _____ _____

3.

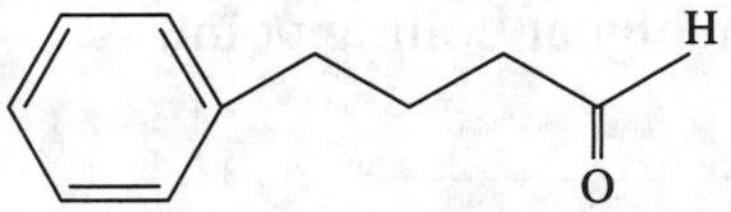

_____ _____ _____

4.

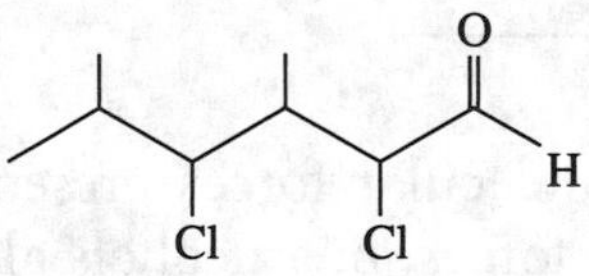

_____ _____ _____

5.

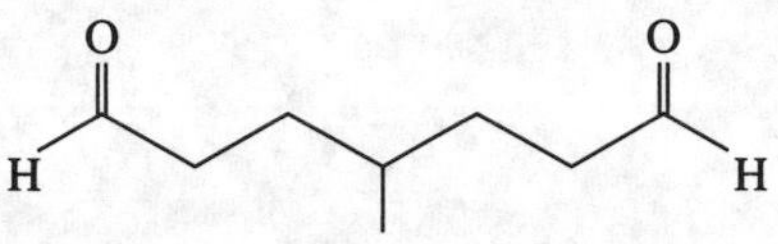

_____ _____ _____

Properties of Aldehydes

The main intermolecular forces present in the liquid and solid forms of aldehydes are dipole-dipole intermolecular forces (IMFs), although dispersion forces are also present. As a consequence of these forces, most aldehydes are liquids at room temperature (22°C). The exceptions are the aldehydes with lighter mass, such as methanal and ethanal, which are gases at room temperature.

	Melting point	Boiling point
Methanal	–92°C	–21°C
Ethanal	–123.5°C	20.2°C
Propanal	–81°C	49°C
Butanal	–96.8°C	74.8°C
Pentanal	–76°C	103°C

As the length of the chain increases, the melting point of the plain aldehydes generally increases, although there is an observed difference between compounds with an odd number of carbons and an even number of carbons as shown above. This effect is found with comparison of the melting points of most organic compounds based on hydrocarbon chains and has to do with how well the chains pack together in the solid. In contrast, as the length of the carbon chain increases, the boiling point of these aldehydes *always* increases.

Comparing the boiling point of the same number of carbons in a chain of alcohols to that in a chain of aldehydes reveals that alcohols in general, have the higher boiling point.

	Melting point	Boiling point
Methanal	–92°C	–21°C
Methanol	–97.6°C	64.7°C
Ethanal	–123.5°C	20.2°C
Ethanol	–114°C	78.37°C

Why? Remember that boiling points depend on the intermolecular forces present. Both aldehydes and alcohols have dispersion forces and dipole-dipole forces, but alcohols also have hydrogen bonding between molecules, which is a stronger intermolecular force.

Comparing melting points, we see that alcohols appear to have a lower melting point than similar carbon chain aldehydes.

Given the name of an aldehyde and the name of a corresponding carbon chain compound, identify the major intermolecular force(s) present in each compound, predict which will have the higher boiling point, and explain why.

1. butanal and butanol

_____

_____

2. pentanal and pentanol

_____

_____

3. hexanal and hexanol

_____

_____

4. propanal and propane

_____

_____

5. butanal and 2-butene

_____

_____

Low-mass aldehydes, with four or fewer carbons in the chain, are soluble in water. The oxygen in the carbonyl group of aldehydes is highly electronegative and can form hydrogen bonds with the partial-positive-charged hydrogen in water, even though as stated above, H-bonds are unlikely between two aldehyde molecules.

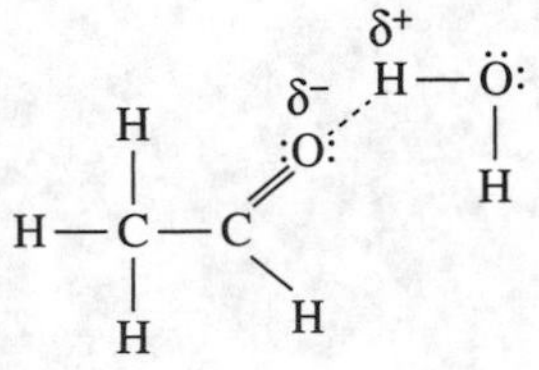

As carbons are added to the chain, the aldehyde becomes less soluble in water because it more closely resembles a hydrocarbon. The majority of a long carbon chain aldehyde is nonpolar, so it is not soluble in water.

Ketones

7

Ketones are a class of organic compounds that have the general structural formula

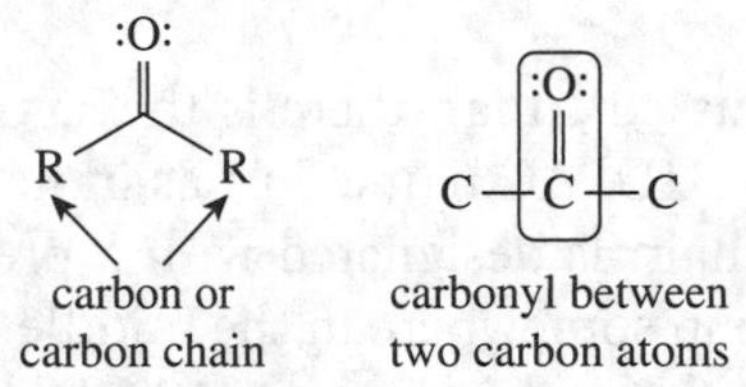

The main structural feature of a ketone is a carbonyl group, a carbon double-bonded to oxygen. This is the same group as in an aldehyde, except that in a ketone the carbonyl group is never on the end. Ketones have two carbon atoms attached to the carbonyl group, whereas aldehydes have only one and one hydrogen because it is on a terminal carbon. Thus the carbon in the carbonyl group is *inside* the chain and doesn't look like a substituent. Ketones are less reactive than aldehydes because the carbonyl group is inside the chain, not on a terminal carbon and there is no adjacent C-H bond to react.

Compare the Lewis structure of the ketone above to that of the aldehyde shown below.

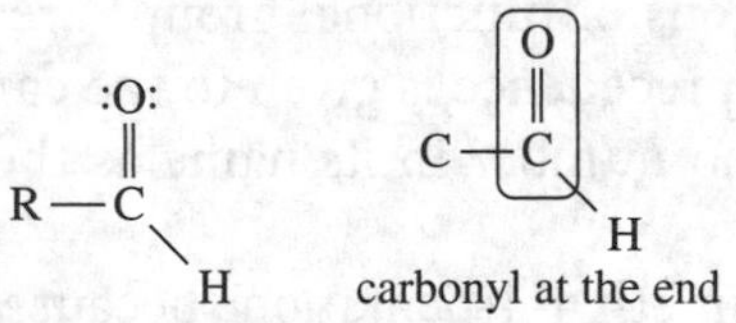

Since the carbon to which the carbonyl is attached must not be terminal, the lowest number of carbons there can be in the chain is three. Therefore, the simplest ketone is propanone. Note that this name follows the usual rules with prefix prop- as there are three C atoms in the longest chain.

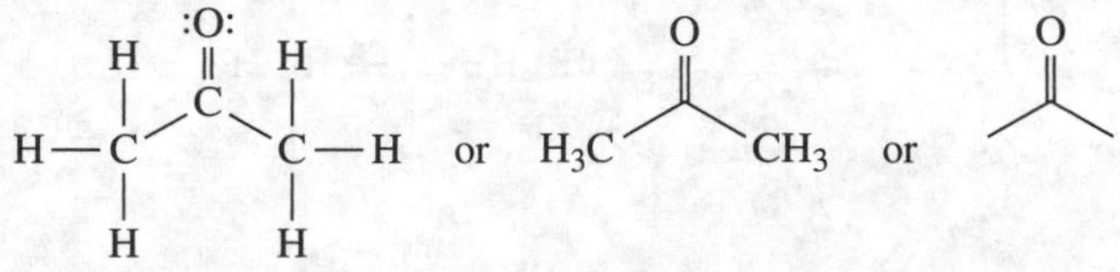

Propanone, like many organic compounds that have been known for a long time, also has a common name. The common name of propanone is acetone, and it is also sometimes called dimethyl ketone. Note that there is a methyl group on each

143

side of the carbonyl group—hence the name. There are many uses for acetone; for instance, it is often used in nail polish remover. Acetone is unique in that it is miscible in water yet can be used as an organic solvent. For this reason, it is a useful cleaning solvent, dissolving both organic and water-based forms of "dirt."

Ketones, which generally have pleasant odors, can be found in many natural and commercial products. The ketones testosterone and progesterone are the male and female sex hormones, respectively. They can be recognized as ketones by the suffix -one in their names. For example, shown below is 2-heptanone which is a component of blue cheese.

As with all organic molecules, discussed so far, there is the same systematic way to name the ketones and each ketone has a unique IUPAC name. Just as in naming alkanes and alkenes, the number of carbons in the longest chain is designated with a prefix. The suffix -one is used to indicate the presence of a ketone group somewhere in the middle of the carbon chain. The chain has to pass through the carbon with the carbonyl group attached, and the carbons in the chain are numbered from the direction that gives the functional group (-one) the lowest number. The carbon atom with the carbonyl group attached is used as the number to identify the location of the ketone. For example, 2-pentanone has the Lewis structure

As is true for the alkanes, the locations of functional groups are noted with the lowest possible number of carbon in the chain, but precedence is given to the carbonyl group having the lowest number. Note that propanone has no number in its name as the only location possible for the carbonyl group is at carbon-2.

3-methyl-2-pentanone (NOT 3-methyl-4-pentanone because of the precedence rule) has the following Lewis structure:

Example: Name the following organic structure.

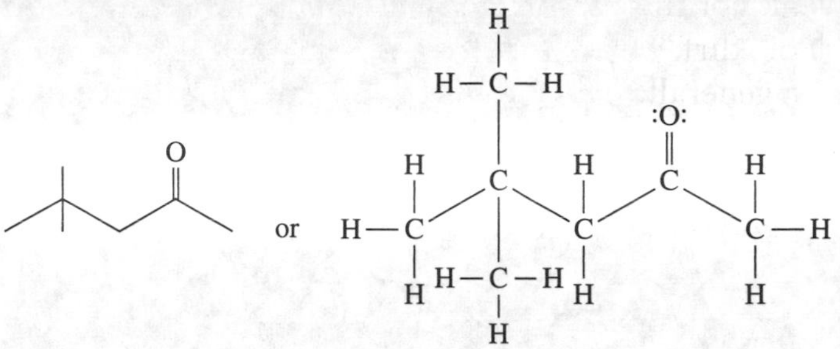

The main chain has five carbons. Numbering from right to left, the carbonyl group is on the second carbon. Two methyl groups are found on the fourth carbon. Hence the compound is named 4,4-dimethyl-2-pentanone. Remember that more than just carbon groups can be substituted on the chain.

EXERCISE

7·1

Given the following structures, name each compound and write its formula.

NAME	FORMULA

1.

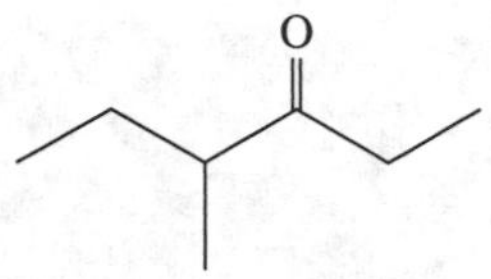

_____   _____

2.

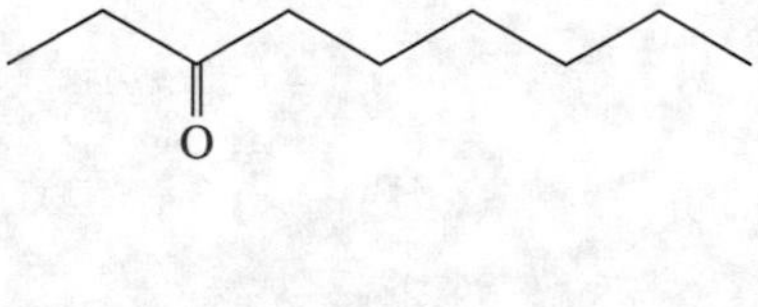

_____   _____

3.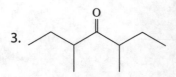

4.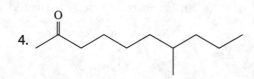

_____ _____

_____ _____

5.

_____ _____

For each of the following compounds, draw the structure and write the formula of the compound.

STRUCTURE   FORMULA

1. 3-chloro-2-pentanone   _____   _____

2. 3-ethyl-2-pentanone   _____   _____

3. 1-bromo-propanone   _____   _____

4. cyclohexanone   _____   _____

5. 1-iodo-2-methyl-3-hexanone   _____   _____

We can represent the structure of ketones in several different ways. It is important to be able to recognize and visualize a molecule using a variety of different representations. These are the same representations used with alkanes and alkenes. The difference is the presence of the carbonyl group. Remember the carbonyl group has a double bond from carbon to the oxygen. For this carbon in the carbonyl group, the hybridization is sp^2 and the shape of the bonds is trigonal planar around this carbon.

For 2-butanone, these representations include

Lewis Structure

Molecular Geometry

Condensed Structural Formula

$CH_3COCH_2CH_3$

Line Drawing

Remember that a "line drawing" shows just the minimum amount of information. If the molecule is a hydrocarbon, the carbon connections are shown, but not the hydrogens. If more than one carbonyl group is in the structure, both are indicated with the suffix -dione.

EXERCISE

7·3

For each of the following compounds, draw the Lewis structure.

1. 3-pentanone

2. 3-methyl-2-butanone

3. 3-ethyl-2-pentanone

4. 2,4-pentanedione

5. 2,4-dimethyl-3-pentanone

EXERCISE

7·4

For each of the following compounds, draw the molecular geometry.

1. 3-pentanone

2. 3-methyl-2-butanone

3. 3-ethyl-2-pentanone

4. 2,4-pentanedione

5. 2,4-dimethyl-3-pentanone

For each of the following compounds, draw the condensed chemical structure.

1. 3-pentanone

2. 3-methyl-2-butanone

3. 3-ethyl-2-pentanone

_____

4. 2,4-pentanedione

_____

5. 2,4-dimethyl-3-pentanone

_____

EXERCISE
7·6

For each of the following compounds, draw the line drawing.

1. 3-pentanone

_____

2. 3-methyl-2-butanone

_____

3. 3-ethyl-2-pentanone

_____

4. 2,4-pentanedione

_____

5. 2,4-dimethyl-3-pentanone

_____

Ketones having a ring of carbons are named as cycloalkanones. The "alka" part is the prefix for the carbon chain. The carbonyl carbon atom is identified with the number 1. The ring is numbered in such a way as to give the lowest numbers of the substituents attached to the ring. For example, 2-methylcyclohexanone has the following structure.

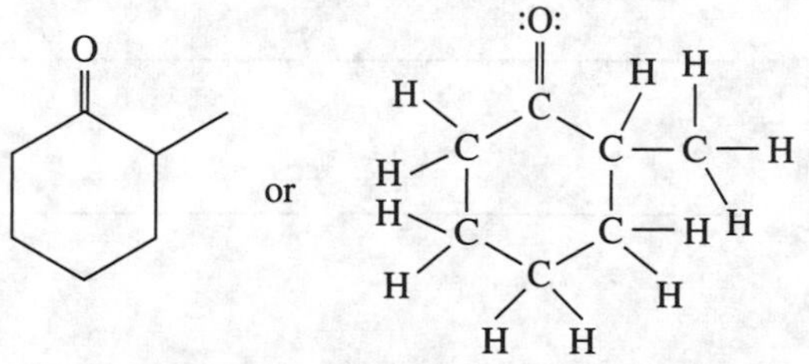

Ketones of another form are those with an added phenyl group. A phenyl group, C_6H_5, receives a higher priority for a substituent when in a ketone structure.

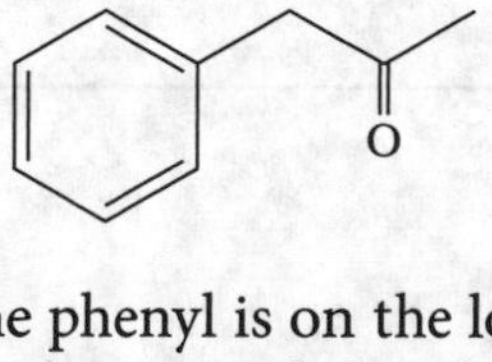

The compound is numbered so that the phenyl is on the lowest-numbered carbon, which would be 1. The carbonyl group is added to carbon 2, and the chain is three carbons long. The name of this compound is 1-phenylpropan-2-one, and it has the common name phenylacetone.

Using the same method, name the following structure.

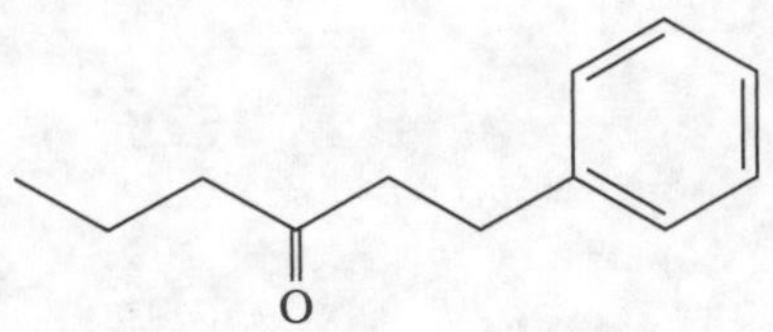

There are seven carbons in the main chain (hept-). The phenyl group should be added to the lowest-numbered carbon, so numbering from right to left, the phenyl would be on carbon 1. Carbon 4 has the carbonyl group. The name of this compound, then, is 1-phenyl-4-heptanone.

Ketone functional groups can also occur in compounds classified as alkenes. The ketone functional group takes priority with respect to the lowest number compared to the double bond of the alkene functional group. In the naming of the compound, the location number of the double bond is given before the longest base carbon chain, and the location of the carbonyl group is given before the suffix -one. In the following example, the longest base chain is six carbons, and the double bond begins at carbon 4. The carbonyl is carbon 2.

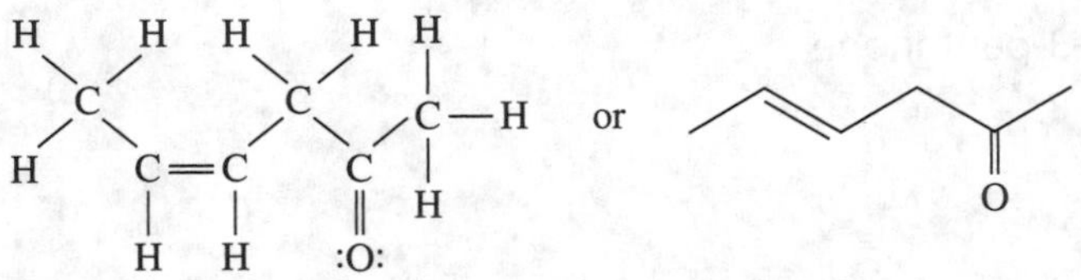

The name of this compound is 4-hexen-2-one. Why not 2-hexene-4-one? Because as stated above, the ketone group takes precedence in over the alkene function group.

EXERCISE

7·7

For each of the following ketones, draw the structure for one isomer, write the name of the compound, and draw its line drawing.

STRUCTURE	NAME	LINE DRAWING

1. $C_3H_7COC_3H_7$ \underline{\hspace{3cm}} \underline{\hspace{3cm}} \underline{\hspace{3cm}}

2. $C_6H_5C_3H_6COCH_3$ \underline{\hspace{3cm}} \underline{\hspace{3cm}} \underline{\hspace{3cm}}

3. $C_3H_5Br_2COCH_2CH_3$ \underline{\hspace{3cm}} \underline{\hspace{3cm}} \underline{\hspace{3cm}}

4. $(C_6H_5)CH_2COCH2CHCl_2$ \underline{\hspace{3cm}} \underline{\hspace{3cm}} \underline{\hspace{3cm}}

For each of the following ketones, write the molecular formula, the name of the compound, and the condensed structural formula.

MOLECULAR FORMULA   NAME   CONDENSED STRUCTURE

1. _____ _____ _____

2. _____ _____ _____

3. _____ _____ _____

4. _____ _____ _____

5. _____ _____ _____

Properties of Ketones

Although dispersion forces are present, the main intermolecular force present in the liquid and solid forms of a ketone are dipole-dipole forces. This results in most ketones being liquids at room temperature and soluble in water. Comparing different organic molecules of the same chain length shows that the types of IMFs make the difference. Pentane has only dispersion forces, whereas 2-pentanone has dipole-dipole interactions. The dipole-dipole interactions will be stronger, so the 2-pentanone will have a higher boiling point. The melting points and boiling points of ketones also increase as the length of the carbon chain increases.

	Melting point	Boiling point
Propanone	−95°C	56°C
2-Butanone	−86°C	80°C
2-Pentanone	−78°C	101°C
2-Hexanone	−57°C	127°C

EXERCISE

7·9

Using your knowledge of the IMFs present in previously covered organic compounds, draw the line drawing for each molecule, indicate which of the following molecules will have the higher boiling point, and explain why.

1. propane or propanone

2. 2-pentene or 2-pentanone

3. 2-butanone or 3-hexanone

4. ethyne or 2-pentanone

5. 2,3-butanedione or butane

Carboxylic Acids

•8•

You are very likely to be familiar with certain organic acids as they are associated with food, beverages, and other substances. Citric acid is found in citrus fruits. Acetic acid is what gives vinegar a sharp taste. Lactic acid is a by-product of strenuous exercise. Formic acid is produced by fire ants. The distinguishing structural feature of all carboxylic acids is the carboxyl group.

carboxyl group	carboxylic acid	condensed structure
$-\overset{\ddot{O}}{\underset{\underset{H}{\ddot{O}}}{C}}$ or $-\overset{O}{\underset{OH}{C}}$	$R-\overset{O}{\underset{OH}{C}}$	$R-COOH$

example	example condensed
$\overset{H}{\underset{H}{H-C}}-\overset{\ddot{O}}{\underset{OH}{C}}$ ethanoic acid or (commonly) acetic acid	CH_3COOH

In a previous chapter, we studied a class of organic compounds called alkanes that had carbon-carbon single bonds in the main carbon chain and found that the length of the carbon chains is a feature used to help name organic molecules. We will make use of this concept again in naming carboxylic acids.

As with all other organic compounds studied so far, the IUPAC name of a carboxylic acid is based on counting the longest carbon chain. The alkane associated with the number of carbons in the longest chain, which includes the carboxyl group, is used as the root. The "e" from the alkane is dropped and replaced with -oic and then followed by the word "acid." In numbering the carbon chain, start with the carbon in the carboxyl group to indicate the location of the carboxyl group. Since the carboxyl group must be located at the end of a carbon chain as was the aldehyde group, this carbon is taken to be carbon 1.

$CH_3CH_2CH_3$ propane
3   2   1

CH_3CH_2COOH propanoic acid
3   2   1

159

The steps in correctly name a simple carboxylic acid are as follows:

- Identify the longest chain including the carboxyl carbon, and write the name of the alkane corresponding to the longest carbon chain.
- The "e" from the alkane is dropped and replaced with -oic, and then followed by the word "acid."
- If another functional or branched group is present on one of the carbons in the carbon chain, identify the carbon in the chain, and number the group with the carbon number.

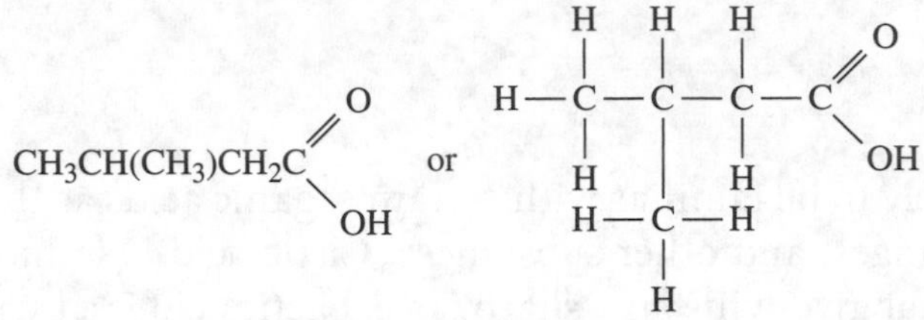

In the example above, counting the carbon atoms including the carboxyl group (right to left), a methyl group is located on the third carbon. The length of the chain is four carbons, and the functional group is a carboxyl. Hence, the name of this compound is 3-methyl-butanoic acid.

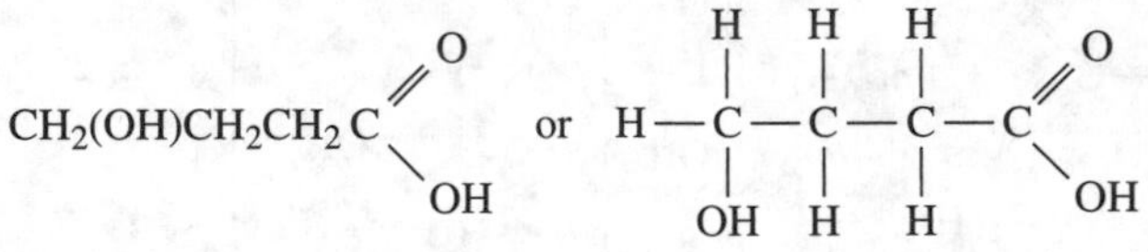

The compound above is an acid (there is a carboxyl group), and it also has a hydroxyl substituted on the chain. Counting the carbons from right to left, on the fourth carbon is a hydroxyl group. There are four carbons in the chain with a carboxyl as the functional group. The name of this compound is 4-hydroxy-butanoic acid. Note that this is not classified as an alcohol in spite of the presence of the –OH group.

3-Methyl-butanoic acid and 4-hydroxy-butanoic acid are examples of aliphatic acids. Aliphatic acids have an open structure where the chains are linked so that alkyl groups are bonded to the carboxyl group. In contrast, nonaliphatic acids, which are also called aromatic acids, have a ring structure where an aryl group is bonded to the carboxylic acid. Aryl is the general name for groups like phenyl- ring structures that we have seen before. Consider the following compound:

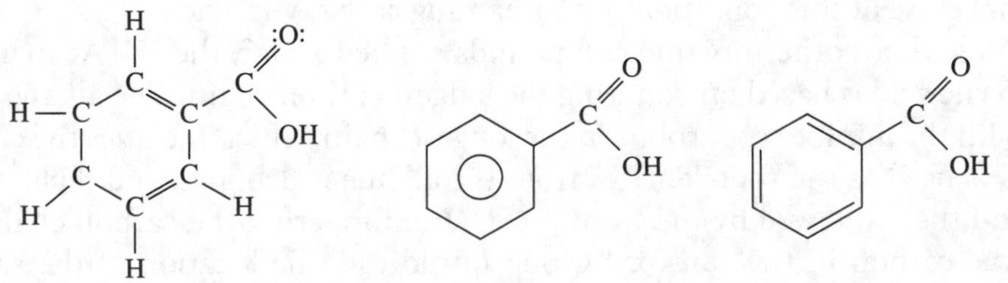

This compound, C_6H_5-COOH, is an example of an aromatic acid. It is known as benzoic acid, but it may also be called benzenecarboxylic acid.

Remember that the ring can be represented with the circle inside the carbon frame as in benzene. This represents the alternating double bonds.

Consider 3-napthoic acid:

This compound, $C_6H_4C_4H_3COOH$, can also be called 3-napthalenecarboxylic acid.

For some carboxylic acids, common names that were in use before the IUPAC nomenclature system was developed can also be used.

Formula	Common name	IUPAC name
CH_3COOH	acetic acid	ethanoic acid
CH_3CH_2COOH	propionic acid	propanoic acid
$CH_3CH_2CH_2COOH$	butyric acid	butanoic acid
$CH_3CH_2CH_2CH_2CH_2COOH$	caproic acid	hexanoic acid

In using common names, the positions of groups attached to the longest continuous carbon chain are designated using letters from the Greek alphabet: alpha (α), beta (β), gamma (γ), delta (Δ), and so on, starting from the first carbon in the carboxylic acid carbon.

Remember that in this case, the first carbon is counted as the one in the -COOH group. Let's name the following compound:

Find the longest chain including the C in the carboxyl functional group (don't forget chains can bend). The longest chain has six carbons, so the end of the name is hexanoic acid. Number the carbons from right to left, and identify the carbons to which the additional groups are attached. The first carbon has a *bromo-* group, the second carbon a *chloro-* group, and the third carbon a *methyl-* group. Hence, the name of this compound is α-bromo-β-chloro-γ-methyl-caproic acid or α-bromo-β-chloro-γ-methyl-hexanoic acid.

The carboxylic acid group has priority in naming over alcohols, aldehyde, and ketones. The prefix -oxo is used to identify an aldehyde or ketone group and the number of the carbon in the longest carbon chain. "*n*"-oxo- is used to designate the location of the -oxo- group, where *n* is the carbon number of the -oxo- group. For example, to name the following compound

$$H-\overset{\overset{\displaystyle H}{|}}{\underset{\underset{\displaystyle H}{|}}{C}}-\overset{\overset{\displaystyle O}{||}}{C}-\overset{\overset{\displaystyle H}{|}}{\underset{\underset{\displaystyle H}{|}}{C}}-\overset{\overset{\displaystyle H}{|}}{\underset{\underset{\displaystyle Cl}{|}}{C}}-C\overset{\displaystyle O}{\underset{\displaystyle OH}{\diagup}}$$

First number the carbons.

$$\underset{5}{C}-\underset{4}{C}-\underset{3}{C}-\underset{2}{C}-\underset{1}{C}\overset{\displaystyle O}{\underset{\displaystyle OH}{\diagup}}$$

The length of the chain is five carbons. The additional groups are on the second carbon, a chloro-, and the fourth carbon, a ketone. This compound is called 2-chloro-4-oxo-pentanoic acid.

A dicarboxylic acid has two carboxyl groups and can be named using the suffix -dioic. Many dicarboxylic acids have common names.

Formula	Common name	IUPAC name
HOOC-COOH	ethanedioic	oxalic acid
HOOC-CH$_2$-COOH	propanedioic	malonic acid
HOOC-CH$_2$-CH$_2$-COOH	butanedioic	succinic acid

If additional atoms or groups are present, in this case "alpha" and "beta" are used instead of "1" and "2."

$$\overset{\displaystyle O}{\underset{\displaystyle HO}{\diagdown}}C-\overset{\overset{\displaystyle H}{|}}{\underset{\underset{\displaystyle Br}{|}}{C}}-\overset{\overset{\displaystyle H}{|}}{\underset{\underset{\displaystyle H}{|}}{C}}-C\overset{\displaystyle O}{\underset{\displaystyle OH}{\diagup}}$$

There are two carboxylic groups, one on each end. There are four carbons, and since there is a carboxylic group on each end, we number in such a way as to get the bromine on the lowest number. Because alpha and beta are used, only the carbons between are "numbered." The bromine is on the alpha carbon.

$$\overset{\displaystyle O}{\underset{\displaystyle H}{\diagdown}}C-\overset{\overset{\displaystyle 1st}{\overset{\displaystyle \alpha}{}}}{C}-\overset{\overset{\displaystyle 2nd}{\overset{\displaystyle \beta}{}}}{C}-C\overset{\displaystyle O}{\underset{\displaystyle H}{\diagup}}$$

The name of the compound is α-bromo-succinic acid.

Name the compound represented by each of the following structures.

1.

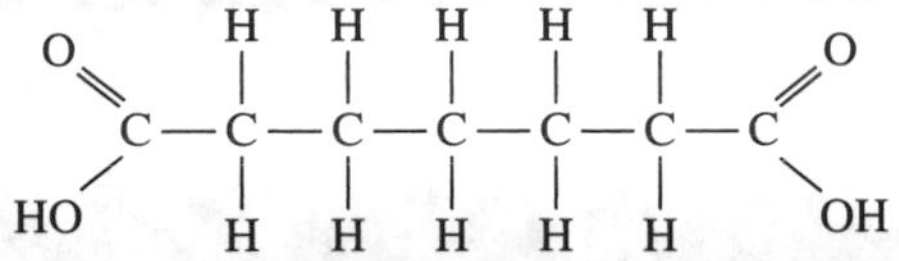

2.

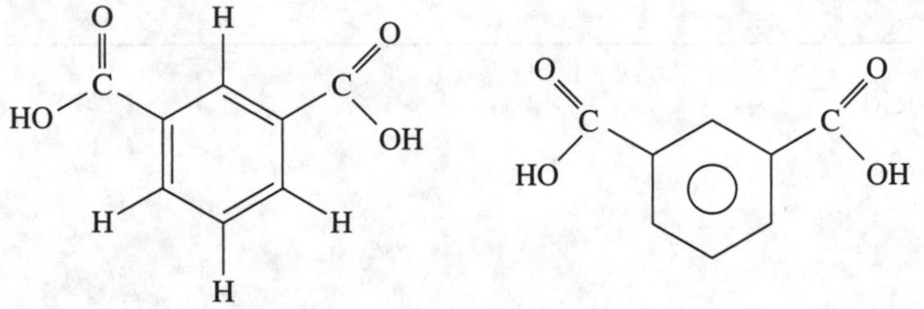

3.

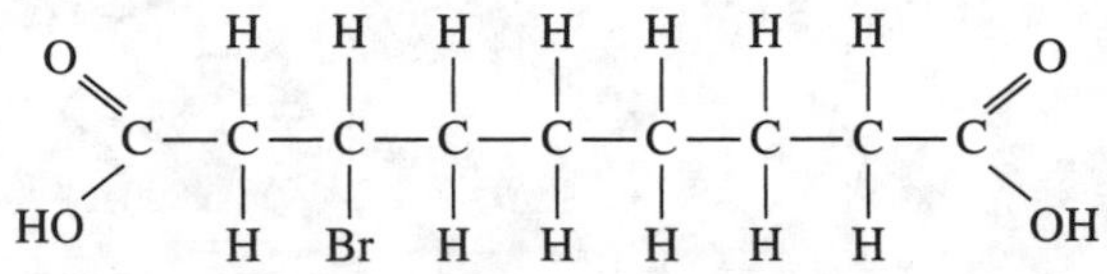

4.

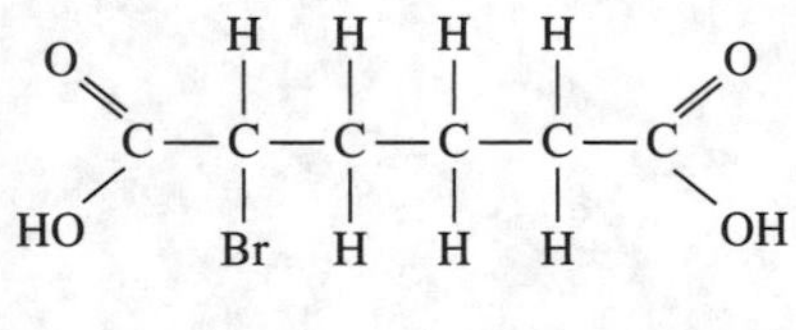

5.

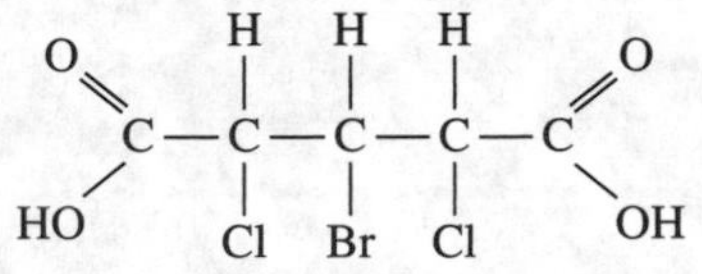

Given the name of a compound, draw the structure of the compound that is known by each of the following names.

1. α-chloro-β-bromo-pentanoic acid

2. 3-hydroxy-2-methylpentanoic acid

3. 4-oxo-2-methylpentanoic acid

4. 4-chloro-4-methyl-2-pentanoic acid

5. α-chloro-pentanedioic acid

EXERCISE

8·3

For each of the following compounds, draw the condensed structural formula and the line drawing.

	CONDENSED STRUCTURE	LINE DRAWING
1. 2-chlorohexanoic acid	______________	______________
2. propanedioic acid	______________	______________
3. 3-hydroxy-5-methyhepatanoic acid	______________	______________
4. β-chloro-butanedioic acid	______________	______________

Physical Properties

Carboxylic acids have high boiling points because of the ability of two carboxylic acid molecules to form dimers. A dimer is a structure that consists of two similar units joined by covalent bonds or by intermolecular forces. In a small sample of acetic acid there are billions of acetic acid molecules involved with hydrogen bonding from the formation of dimers and hydrogen bonding through the interaction of hydrogen bonding with water molecules and acetic acid molecules. In carboxylic acids, hydrogen bonds form.

Acetic acid has the hydrogen attached to an oxygen atom. The oxygen atom's electronegativity is high enough to cause the hydrogen to develop a partial positive charge. The oxygen of the carbonyl group has a partial negative charge. The partial positive charge on the hydrogen of one molecule forms a hydrogen bond with the partial negative charge on the oxygen of another molecule. Because of the close spacing, the partial negative charge on the carbonyl oxygen atom of the second molecule forms a hydrogen bond with the partial positive charge on the hydrogen of the first molecule. These two hydrogen bonds keep the two molecules together as a unit in the liquid phase and even during boiling in the gas phase. The boiling point of acetic acid is higher than expected thanks to the increase in strength of the intermolecular force of attraction between the molecules and the increase in molecular weight of the formed dimer.

The boiling point trend of carboxylic acids (see Table 2) is due to a combination of the strength of intermolecular forces of attraction betweeen molecules and the molecular weight of the molecules. It takes additional energy to increase the kinetic energy of heavier molecules than to increase that of lighter molecules, and it takes additional energy to break the intermolecular forces of attraction.

Table 2 Boiling Point of Three Carboxylic Acids

Carboxilic acid	Boiling point (°C)
methanoic acid	101
ethanoic acid	118
propanoic acid	141

Draw a Lewis structure showing the location of hydrogen bonding between two pentanoic acid molecules that form a dimer, and predict the boiling point for pentanoic acid.

Explain why dimers have higher boiling points than the single molecules they are made from.

Carboxylic acids have sharp, unpleasant odors. For example, caproic acid has the odor of the armpit of a goat. Butanoic acid is excreted by rancid butter. Acetic acid mixed with water (vinegar) has a sharp taste.

Carboxylic acids, in general, are weak acids. In a reaction, an acid is a proton donor. The proton is a hydrogen ion. With the loss of an electron, the hydrogen atom, which consists of one proton and one electron, becomes an ion consisting of only a proton. Not just any hydrogen in the organic acid can be donated in the reaction. The C-H bond is too strong, whereas the O-H bond is weaker. Only the H of the OH can be donated in the reaction; in other words, the acidic proton is the hydrogen atom bonded to the oxygen atom in the carboxylic group that does not have the double bond in the Lewis structure. The hydrogen atoms bonded directly to carbon atoms are not acidic hydrogen atoms. The acidic hydrogen is circled in the following Lewis structure of acetic acid.

$$
\begin{array}{c}
\text{H} \\
| \\
\text{H}-\text{C}-\text{C} \overset{\displaystyle O}{\underset{\displaystyle O\text{(H)}}{}} \quad \text{the H that is donated} \\
| \\
\text{H}
\end{array}
$$

Given the Lewis structure of pentanoic acid, circle the acidic hydrogen atom.

$$\begin{array}{cccccc} & H & H & H & H & \\ & | & | & | & | & \nearrow O \\ H- & C- & C- & C- & C- & C \\ & | & | & | & | & \searrow \\ & H & H & H & H & OH \end{array}$$

A carboxylic acid is a compound that, when placed in water, increases the hydronium ion concentration. A hydronium ion is a water molecule that has accepted a hydrogen ion, forming H_3O^+. The hydrogen ion has been donated by the carboxylic group of the acid.

$$H-\ddot{O}: \quad + \quad H^+ \quad \longrightarrow \quad H-\ddot{O}-H^+$$
$$\quad | \qquad\qquad\qquad\qquad\qquad | $$
$$\quad H \qquad\qquad\qquad\qquad\qquad H$$

water + hydrogen ion $\longrightarrow$ hydronium

pH is a measure of the concentration of hydronium ions in a solution. The pH scale is used to classify aqueous solutions as acidic, basic, or neutral. A solution that tests acidic has lots of hydronium ions. To determine the actual pH, it is necessary to know the concentration of hydronium ions. The formula to find the pH is

$$pH = -\log[H_3O^+]$$

As this formula indicates, the more hydronium ions present, the lower the pH. A lower pH indicates a more acidic solution.

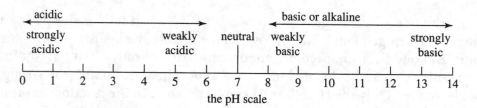

the pH scale

When the reaction occurs, the water molecule forms a hydronium, and a carboxylate ion is left after the removal of the hydrogen ion.

$$\begin{array}{cc} H & \ddot{O} \\ | & \parallel \\ H-C-C & + \\ | & \\ H & \ddot{O}H \end{array} \quad H-\ddot{O}: \quad \rightarrow \quad \left[H-\ddot{O}-H \right]^+ \quad + \quad \left[\begin{array}{cc} H & \ddot{O} \\ | & \parallel \\ H-C-C \\ | & \\ H & \ddot{O}: \end{array} \right]^-$$

acetic acid   water   hydronium ion   acetate ion

The reason why carboxylic acids are considered weak is that in a water solution, not all the acid molecules react to donate hydrogen ions to the water. In fact, very few do. In technical terms, most carboxylic acids are weakly acidic, because in aqueous solutions the acid stays largely intact as a molecule. The proportion of molecules that actually undergo the reaction with water is called the percent dissociation. The percent dissociation of organic acids is usually between 1% and 15%. This means that in a dilute solution of an organic acid in water, the solution is mostly water molecules, a few hydronium ions, and a few more intact acid molecules.

 When 100 H_2O molecules and 10 weak acid molecules (HA) are combined, 1 HA dissociates, forming 1 H_3O^+. Here, 99 H_2O molecules are left out to focus on the HA. H_2O is the main molecule present.

Since the pH of a solution is a measure of the hydronium ion concentration, the fewer the H_3O^+ ions in solution, the less acidic the solution. Therefore, calculating the concentration of hydronium determines the pH. The concentration of the acid (HA) itself can't be used, since only a few of the molecules actually form hydronium. When a carboxylic acid is placed in water, it forms an acidic solution, but more important, it establishes an equilibrium. Establishment of an equilibrium means that the reaction moves forward to donate H^+ and dissociation begins, but then some of the product ions start recombining and reforming HA molecules. At equilibrium, the rate of the forward reaction equals the rate of the reverse reaction, and the number of molecules of HA and hydronium ions remains constant. The equilibrium constant can be written

(a) $HA(aq) + H_2O(l) \rightleftharpoons H_3O^+(aq) + A^-(aq)$

reaction equation
where HA represents any weak acid

(b) equilibrium expression for any weak acid

$$K_a = \frac{[H_3O^+]\,[A^-]}{[HA]}$$

equilibrium constant   [ ] stands for concentration in molarity (moles per liter of solution)

For instance, propanoic acid has an equilibrium constant of 1.8×10^{-5}. The reaction is

$$CH_3CH_2COOH + H_2O \rightleftharpoons CH_3CH_2COO^- + H_3O^+ \qquad K_a = 1.8 \times 10^{-5}$$

The expression is

$$K_a = \frac{[H_3O^+][CH_3CH_2COO^-]}{[CH_3CH_2COOH]}$$

Assuming a 0.10 M solution, and after number substitution, the equation is

$$1.8 \times 10^{-5} = \frac{(X)(X)}{[0.10M - X]}; \quad 1.8 \times 10^{-5} = \frac{X^2}{0.10 - X}$$

$$X = 1.3 \times 10^{-3} M$$

This means that the concentrations at equilibrium are 0.0013 M H_3O^+, 0.0013 M $CH_3CH_2COO^-$, and 0.10 M CH_3CH_2COOH. Thus, in a 0.10 M solution of propanoic acid, the majority of acid particles present in the solution are propanoic acid *molecules*. Note that so few have reacted that we can say the concentration of dissociated propanoic acid is still 0.10 M. Approximately 1.3% of the propanoic molecules react with water to generate hydronium ions and propanate ions. If we initially placed only 100 propanoic acid molecules in a very small volume of water, only one molecule would have reacted with the water.

Taking the concentration of the hydronium ion (x) and substituting it into the pH equation

$$pH = -\log [H_3O^+]$$

$$pH = -\log [0.0013 \ M \ H_3O^+]$$

$$pH = 2.90$$

> The larger the value of K_a, the more product species are present at equilibrium.

Let's compare the pH of 0.10 M propanoic acid to the pH of 0.10 M formic acid, HCCOH. Formic acid and propanoic acid are both weak acids, but have different K_a values.

$$HCOOH + H_2O \rightleftharpoons HCOO^- + H_3O^+ \qquad K_a = 1.7 \times 10^{-4}$$

Note that the K_a for formic acid is larger than the K_a for propanoic acid. Which solution is more acidic or has the lowest pH?

Assuming a 0.10 M solution of formic acid, substituting into the equilibrium expression and solving, at equilibrium we have 0.10 M formic acid, approximately 4.1×10^{-3} M HCOO$^-$ and approximately 4.1×10^{-3} M H$_3$O$^+$. The pH of this solution is about 2.39. As expected, 0.10 M formic acid is a bit more acidic than 0.10 M propanoic acid. This is true because the K_a value for formic acid is a bit larger than the K_a value for propanoic acid.

The magnitude of the equilibrium constant indicates the extent to which the reactants form the products, or, in other words, form an acidic solution. When two different acids of the same concentration are compared, the K_a values indicate which of the two is the stronger acid.

Table 3 Acid Dissociation Constant of Three Carboxylic Acids

	K_a
methanoic acid	1.8×10^{-4}
benzoic acid	6.4×10^{-5}
ethanoic acid	1.8×10^{-5}
propanoic acid	1.3×10^{-5}
butanoic acid	1.5×10^{-5}

EXERCISE
8·7

Of the carboxylic acids listed in Table 3, choose the acid that generates the most hydronium ions when placed in water. Write the reaction equation and equilibrium expression.

EXERCISE

8·8

Assume a 0.10 M concentration of the acid chosen in Exercise 8-7 and solve the equilibrium constant expression for the concentration of the hydronium ion.

EXERCISE

8·9

Using the concentration of hydronium in Exercise 8-8, solve the pH equation for the pH of the acid.

EXERCISE

8·10

Of the carboxylic acids listed in Table 3, choose the acid that generates the fewest hydronium ions when placed in water. Write the reaction equation and equilibrium expression.

EXERCISE

8·11

Assume a 0.10 M concentration of the acid chosen in Exercise 8-10, and solve the equilibrium expression for the concentration of the hydronium ion.

The percent ionized is a measure of how many of the molecules actually dissociated.

$$CH_3CH_2COOH + H_2O \rightleftharpoons CH_3CH_2COO^- + H_3O^+ \qquad K_a = 1.8 \times 10^{-5}$$

The concentration of hydronium ions is 0.0013 M in a 0.10 M propanoic acid solution.

$$\% \text{ ionized} = \frac{[H_3O^+]}{[HA]} \times 100; \quad \% \text{ ionized} = \frac{[0.0013\ M]}{[0.10\ M]} \times 100\%$$

$$\% \text{ ionized} = 1.3\%$$

This means that about one molecule for every 100 molecules dissociates. In contrast, strong acids dissociate nearly 100%. (Organic acids are never strong acids.)

Ethanoic acid is also called acetic acid. Distilled or white vinegar is about 5% acetic acid. White vinegar is used in cooking, baking, preserving foods, and pickling. Some glass cleaners contain acetic acid. When 100 molecules of acetic acid are placed in water, only 5% react. To help keep track of what is happening to each substance, we can set up an ICE table, where I stands for "initial," C stands for "change," and E stands for "equilibrium."

$$CH_3CH_2COOH + H_2O \rightleftharpoons CH_3CH_2COO^- + H_3O^+ \qquad K_a = 1.8 \times 10^{-5}$$

	CH₃CH₂COOH	**CH₃CH₂COO⁻**	**H₃O⁺**
Initial	100 molecules	0	0
Change	−5 molecules	+5 ions	+5 ions
Equilibrium	95 molecules	5 ions	5 ions

% dissociation = number of H₃O⁺ ions at equilibrium/initial molecules of acid × 100% = 5%

$$\% \text{ dissociation} = \frac{5}{100} \times 100 = 5\%$$

This indicates that the extent of dissociation depends on the initial concentration of the acid.

Given an initial count of 200 butanoic acid molecules, if a 0.001 M solution of butanoic acid is 4% dissociated, how many butanoic acid molecules remain at equilibrium? Write a reaction equation, show a calculation for how many molecules dissociate, and set up an ICE table showing how many relative molecules are present at equilibrium.

Benzoic acid, which occurs in some plants, can be used to synthesize other compounds, such as esters. The anion is called benzoate. Sodium benzoate is used as a food preservative. Refer to the K_a values shown in Table 3, and calculate the pH of the solution and the percent dissociation in a 0.10 M solution.

Amines

Organic compounds contain a variety of elements, but the four most common elements are carbon, hydrogen, oxygen, and nitrogen. Organic compounds containing nitrogen are very different from organic compounds containing oxygen. One class of organic compounds containing nitrogen is the amines. Amines have the general formula RNH_2, where the R represents a carbon chain and the $-NH_2$ is called a primary amine group.

$$R-\overset{\displaystyle H}{\underset{\displaystyle H}{N:}}$$

Nitrogen is a key element in DNA and RNA. Caffeine and nicotine contain several nitrogen atoms arranged in an amine structure.

caffeine   nicotine

The simplest amine is methylamine, CH_3NH_2. Methylamine can be thought of as being derived from an ammonia molecule, NH_3, by substituting a methyl group for one of the hydrogen atoms. Methylamine has the following Lewis structure.

ammonia   methylamine

Note that there is one lone electron pair on the nitrogen atom on both ammonia and methylamine. This lone pair causes repulsion of the N-H bonds away from the lone pair, reducing the bond angles. The H-N-H bond angle is approximately 107°. The hydrogen atoms around the central N atom in ammonia form a trigonal pyramidal geometry.

In spite of the differences in nitrogen-containing groups in organic compounds, the naming system follows the same rules as for other substituted molecules we have studied earlier.

First, if there is an -NH$_2$ group it will be at one end of the carbon chain, which is named with the usual prefix for the length of the carbon chain. Then change the ending to -yl, just as you did before when making a substituent group. Then add "-amine" to the name.

The parent methane becomes methyl, and the complete amine name is methylamine.

methane   methyl   methylamine

For each compound, draw the parent molecule (given), and then substitute an -NH$_2$ functional group for the terminal hydrogen to produce an amine.

	PARENT DRAWING	AMINE DRAWING
1. ethane	_____	_____
2. propane	_____	_____
3. butane	_____	_____
4. pentane	_____	_____
5. octane	_____	_____

Name the amines that you drew in Exercise 9-1.

1. _____

2. _____

3. _____

4. _____

5. _____

Not all amines have the $-NH_2$ group. Those that do are termed *primary* amines. If two carbon atom (alkyl) groups are attached to the nitrogen, this is a secondary amine. If three carbon atoms are attached to the nitrogen, then this is a tertiary amine. This is explained in detail in Types of Amines.

When carbon groups are substituted for the H on the NH_2 group, unsymmetrical substituted secondary and tertiary amines use the symbol "N" to indicate the carbon group is bonded to the nitrogen. For example, N,N-dimethylpropylamine has two methyl groups and one propyl group bonded to the nitrogen.

A lowercase "n" indicates a "normal" straight carbon chain. For example, n-propylamine is a chain of three carbons with NH_2 on one end.

Naming branched amines requires identifying alkyl groups. When there is one complex alkyl group bonded to the nitrogen, and the nitrogen atom is part of the longest carbon chain and not a branch, name the alkyl group followed by -amine. Carbon 1 is always bonded to the nitrogen in the amine. For example, here is 2-methyl-1-butylamine:

Name each of the following compounds.

1. 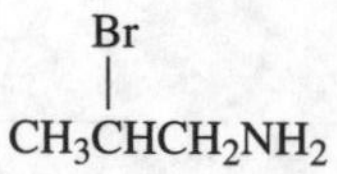   Br
        |
   CH$_3$CHCH$_2$NH$_2$

_____

2. CH$_3$CHCH$_2$Br
        |
       NH$_2$

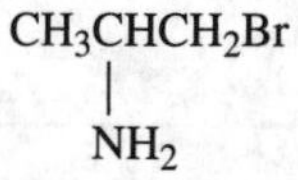

_____

3. 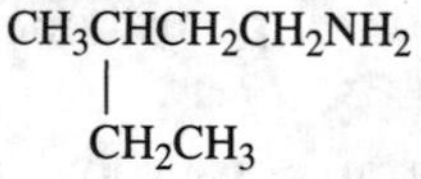 CH$_3$CHCH$_2$CH$_2$NH$_2$
        |
       CH$_2$CH$_3$

_____

4.

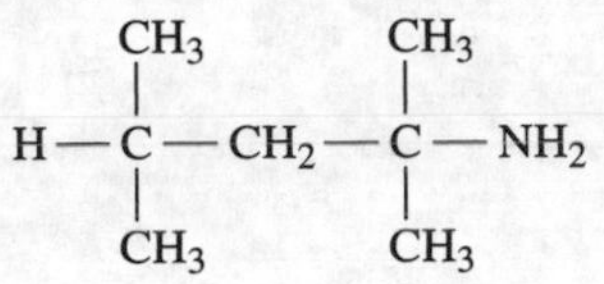

$$\underset{\displaystyle \overset{\displaystyle CH_3}{|}}{\underset{\displaystyle \underset{\displaystyle CH_3}{|}}{H-C}}-CH_2-\underset{\displaystyle \underset{\displaystyle CH_3}{|}}{\overset{\displaystyle \overset{\displaystyle CH_3}{|}}{C}}-NH_2$$

_____

5.

$$H-\underset{\displaystyle \underset{\displaystyle CH_3}{|}}{\overset{\displaystyle \overset{\displaystyle CH_3}{|}}{C}}-\underset{\displaystyle \underset{\displaystyle NH_2}{|}}{\overset{\displaystyle \overset{\displaystyle H}{|}}{C}}-\underset{\displaystyle \underset{\displaystyle CH_3}{|}}{\overset{\displaystyle \overset{\displaystyle CH_3}{|}}{C}}-CH_3$$

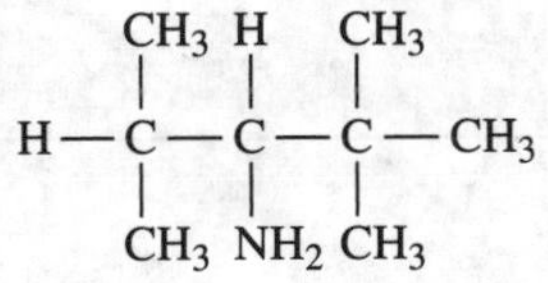

Remember, the symbol "N" is used to designate which alkyl groups are attached to the nitrogen atom in an amine when the nitrogen atom is not a branch of a carbon in the longest carbon chain. The number in front of the parent alkane followed by the phrase "-amine" indicates on which carbon the nitrogen atom is attached. For example, here is N,N-dimethyl-2-butanamine:

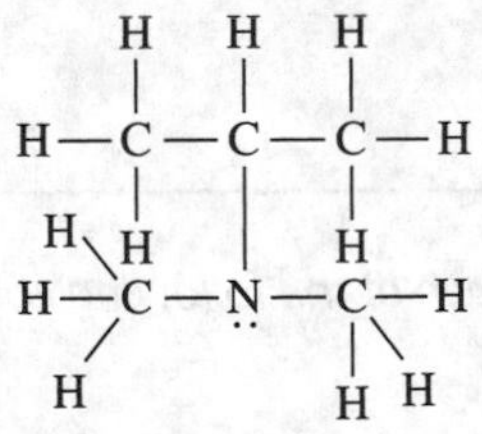

The name "N,N-dimethyl" indicates that both methyl groups are attached to the N, and "-2-" in front of "butanamine" indicates on which carbon atom the nitrogen is attached.

N-methyl-2-methyl-3-pentanamine has the following structure:

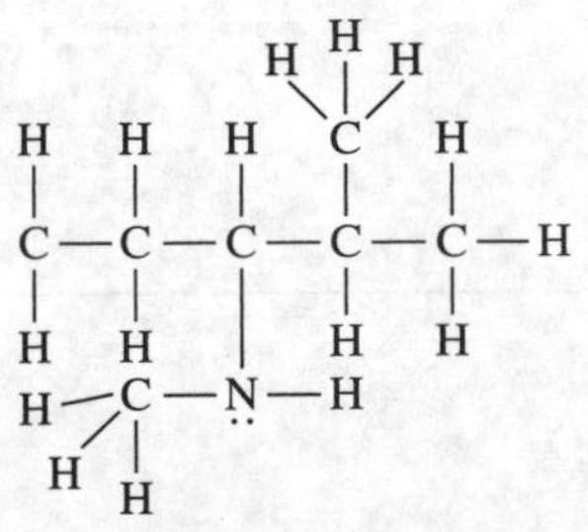

Name each of the following compounds.

1.

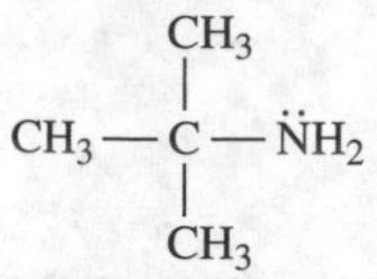

$$CH_3-\overset{\displaystyle CH_3}{\underset{\displaystyle CH_3}{\overset{|}{\underset{|}{C}}}}-\ddot{N}H_2$$

_____

2. $CH_3CH_2-\overset{\displaystyle \ddot{N}-H}{\underset{\displaystyle |}{\underset{\displaystyle CH_3}{|}}}$

_____

3.

$$CH_3-\overset{\displaystyle H}{\underset{\displaystyle H}{\overset{|}{\underset{|}{C}}}}-\overset{\displaystyle :N-CH_3}{\underset{\displaystyle H}{\overset{|}{\underset{|}{C}}}}-CH_2CH_3$$

with CH_3 on the left of N

_____

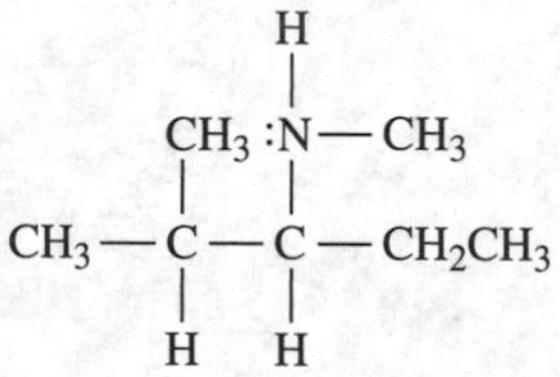

4. 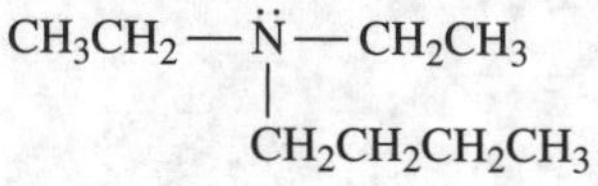 $CH_3CH_2-\overset{\displaystyle \ddot{N}-CH_2CH_3}{\underset{\displaystyle |}{\underset{\displaystyle CH_2CH_2CH_2CH_3}{|}}}$

_____

5.

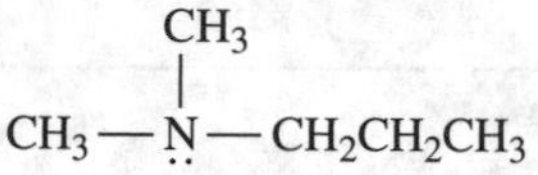

$$CH_3-\overset{\displaystyle CH_3}{\underset{\displaystyle |}{\overset{|}{N}}}-CH_2CH_2CH_3$$

_____

Draw the structure for each of the following compounds.

1. 2-methylcyclopentanamine

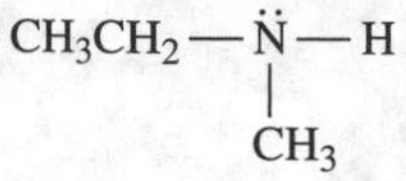

2. 2-phenylethylamine

_____

3. N-ethyl-4 methyl-2-hexylamine

_____

4. triethylamine

_____

5. N-propylhexanamine

_____

If other functional groups are added, the naming of the compound changes because in the order of precedence, amines are last on the list. This means that when the carbon chain has a functional group, the functional group takes precedence, and "amine" is changed to "amino" with a number to indicate where the amine group is located on the carbon chain. For example, here is 3-amino-1-butanol:

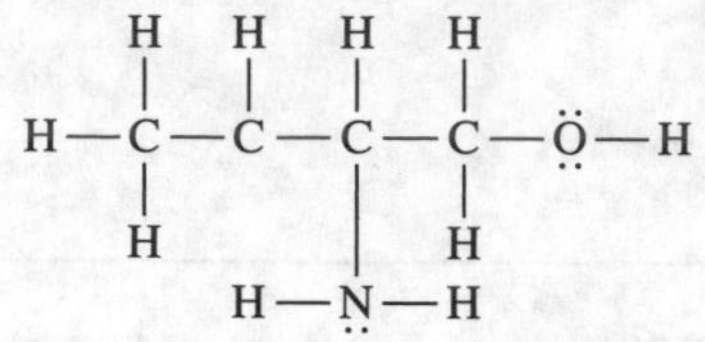

$$
\begin{array}{c}
\quad\ \ H\ \ \ H\ \ \ H\ \ \ H \\
\quad\ \ |\ \ \ \ |\ \ \ \ |\ \ \ \ | \\
H-C-C-C-C-\ddot{O}-H \\
\quad\ \ |\ \ \ \ \ \ \ \ \ \ \ |\ \ \ \ \ \ \ |\\
\quad\ \ H\ \ \ \ \ \ \ \ \ \ \ H \\
\quad\quad\quad\ \ H-N-H
\end{array}
$$

Name each of the following compounds.

1. $H_2\ddot{N}CH_2CH_2CH_2OH$ _____

2. 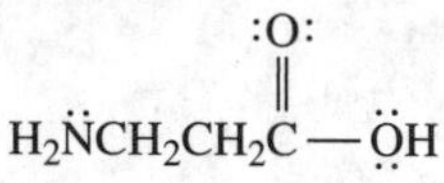
   $$H_2\ddot{N}CH_2CH_2\overset{\overset{\displaystyle :O:}{\|}}{C}-\ddot{O}H$$ _____

3. 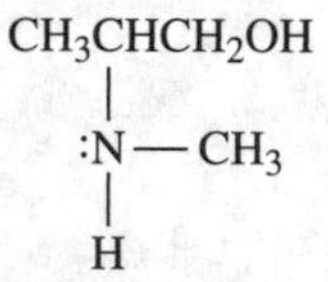 CH_3CHCH_2OH
   $\quad\quad |$
   $\quad\ :N-CH_3$
   $\quad\quad |$
   $\quad\quad H$ _____

4. 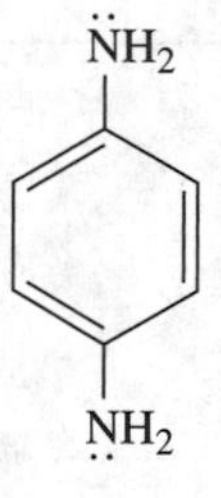 $\ddot{N}H_2$

   (benzene ring)

   $\ddot{N}H_2$ _____

5. 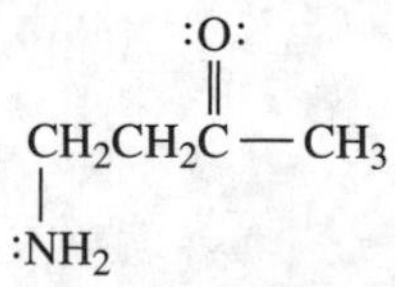
   $$CH_2CH_2\overset{\overset{\displaystyle :O:}{\|}}{C}-CH_3$$
   $\ |$
   $:NH_2$ _____

EXERCISE

9·7

Draw the following:

1. 2-aminoethanol

2. N-propyl-2-methylcyclopentanamime

3. 4-aminopentanoic acid

4. 5-aminopentanal

5. 1-amino-3-pentanone

Amine functional groups can also be found on compounds that are not in straight chains. One group consists of the cyclo-compounds. A compound with an -NH$_2$ group bonded to cyclohexane is called cyclohexanamine or (more commonly) cyclohexylamine.

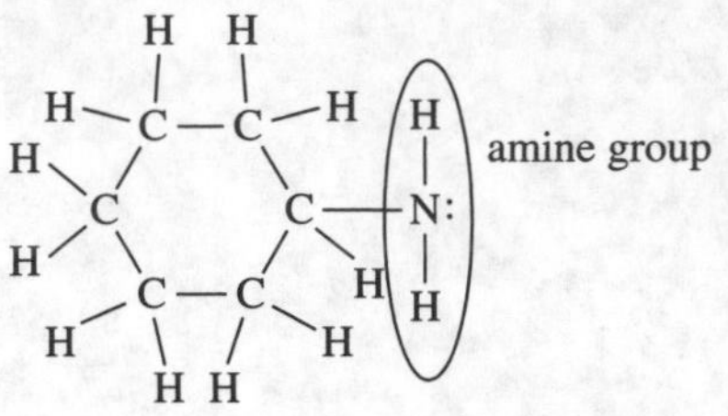

amine group

As with other organic compounds, a phenyl group can be added. The simplest amine containing a phenyl group attached to the –NH$_2$ amine (which may also be called an aryl group) is aniline:

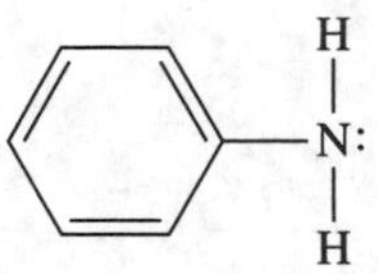

To name compounds with an aniline structure, number the carbon atoms in the ring with carbon 1 being the carbon where the nitrogen atom is attached.

If other groups are added to the ring, remember that you must include the number of the carbon where they are added and that the amino group is always carbon 1. Then the prefixes ortho- (1 and 2), meta- (1 and 3), and para- (1 and 4) are used to indicate where a substituent is located on the ring. For example, in *p*-bromoaniline, carbon 1 has an amino group and carbon 4 has a bromo group.

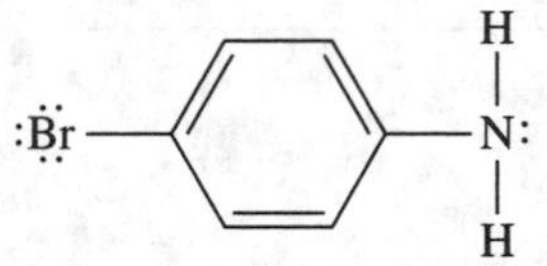

For alkyl groups attached to the nitrogen atom, use the "N" system.

N-ethylanline is a secondary amine (two carbon groups are attached to the nitrogen).

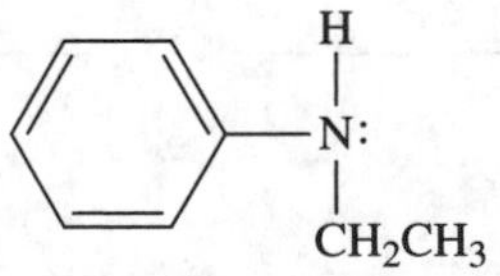

Name each of the following compounds.

1. _____

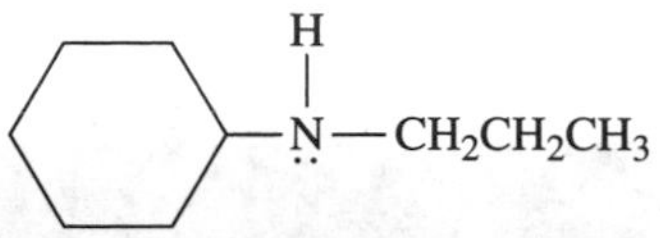

2. _____

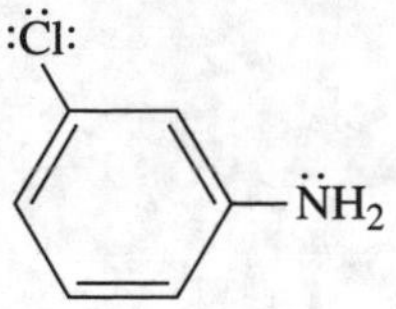

3. 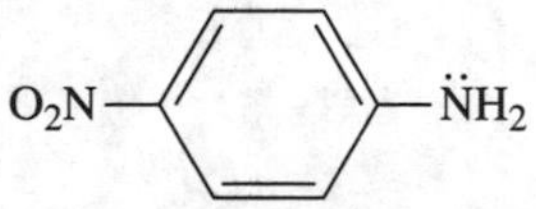 O$_2$N—⟨ ⟩—N̈H$_2$ _____

4. _____

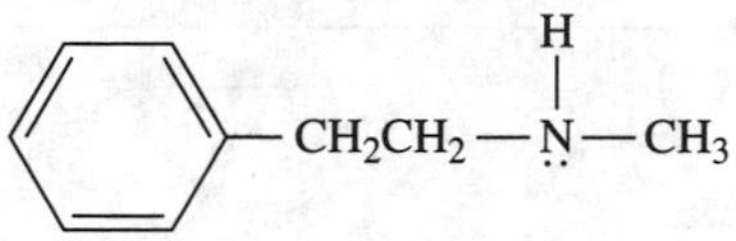

5. _____

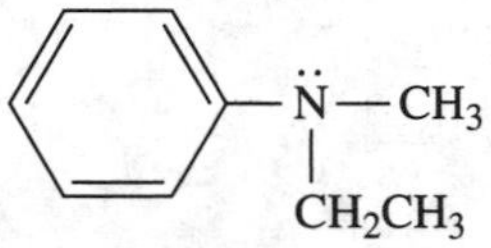

Draw each of the following compounds.

1. N-ethylcyclohexylamine

2. N,N-diethylaniline

_____

3. *o*-ethylaniline

_____

4. *m*-methyl-*p*-aminobenzonitrile

_____

5. *p*-methylaniline

_____

Sometimes an atom of an element that is not carbon can be in a ring of carbon atoms. Such compounds are termed *heterocyclic* compounds. If nitrogen is present, these are methine compounds, which have rather different properties and names than amines. Pyridine and piperidine are examples of heterocyclic -ines. Piperdine is a colorless liquid with a pepper-like odor. The name comes from the Latin word for pepper, "piper." Many of these compounds do not follow a strict naming system, and you just have to learn their names. Piperidine is an example

pyridine   piperine

Piperidine is produced by the reaction of pyridine with hydrogen gas in the presence of a catalyst. Piperidine is used as a starting material for pharmaceuticals. The piperidine structure is found in alkaloids. The closely related compound piperine is what gives black pepper a spicy taste.

Demerol is an analgesic that contains a cyclic amine.

demerol

Some allergy medicines contain amines or amides as the active ingredient. Benadryl is an example of an antihistamine. The amine in the name indicates that there is an amine present. Several common cold or allergy medicines available at a pharmacy contain amines. Bromopheniramine is the active ingredient in some allergy medicines.

bromopheniramine

Two infamous chemicals, methamphetamine and methoxy amphetamine, are illegal drugs that when ingested cause a psychological reaction.

methamphetamine   methoxy amphetamine

Both compounds contain an amine functional group.

Types of Amines

Remember amines can be classified in terms of the number of carbon groups directly attached to the nitrogen atom. Many amines discussed above are called primary (1°) amines. There is only one carbon chain associated with a primary amine. If a second R- group is bonded to nitrogen, the structure is called a secondary amine. In other words, there is a chain on two sides of the amine group. Dimethylamine is a secondary (2°) amine.

1° amine
one carbon attached
to nitrogen

2° amine
two carbons attached
to nitrogen

If there are three alkyl groups attached to the central nitrogen, the compound is classified as a tertiary (3°) amine.

$$H_3C - N: \quad \begin{array}{c} CH_3 \\ | \\ | \\ CH_3 \end{array}$$

When two (or three) identical alkyl groups are attached to the nitrogen, the prefix di- (or tri-) is used in front of the alkyl group. Consider the structure:

3° amine
three carbons attached
to nitrogen

This compound is named trimethylamine because three methyl groups are attached to the amine. Here is another example:

This compound is named triethylamine. It is also sometimes called N,N-diethylethanamine.

Identify each of the following compounds as a primary, secondary, or tertiary amine.

1. $CH_3CH_2\overset{..}{N}H_2$ _____

2. $CH_3 — \overset{..}{N} — CH_2CH_3$
            |
           CH_3 _____

3. $CH_3CH_2 — \overset{..}{N} — H$
              |
              CH_3 _____

4. $\overset{..}{N} — CH_3$ with phenyl and H _____

5. $CH_3 — \overset{..}{N}$ with phenyl and CH_2CH_3 _____

Properties of Amines

One prominent physical characteristic of amines is their disagreeable odor. Light-weight amines can smell like dead fish. Medium-weight amines can smell like well-worn socks. Ammonia and methylamine are both water soluble. Amines containing more than seven carbon atoms are usually not water soluble. This can be explained by the forces they exhibit.

Dipole-dipole intermolecular interactions exist between amine molecules as a consequence of the C-N bond being polar. Primary and secondary amines can exhibit hydrogen bonding, another type of intermolecular force of attraction, because of the highly polar N-H bond. When placed in water, amines containing fewer than seven carbons are water soluble due to the intermolecular forces of attraction between the oxygen of the water molecule and the hydrogen attached to the nitrogen of a primary or secondary amine. Also, the partial positive charge hydrogen from a water molecule is attracted to the partial negative charge of nitrogen atom of the amine. Both of these intermolecular forces of attraction are called hydrogen bonds.

Compared to alkanes with the same number of carbons, amines have higher boiling points.

Table 9-1 Comparison of an Alkane to a Primary Amine of Similar Molecular Weight

	Condensed structure	Molecular weight	Boiling point (°C)
Propane	$CH_3CH_2CH_3$	44.09	−42
Ethylamine	$CH_3CH_2NH_2$	45.09	17

The reason for the difference in boiling points is the type and strength of intermolecular forces that are present. A sample of propane liquid is held together as a liquid by the intermolecular forces known as London dispersion forces. A sample of liquid ethylamine also exhibits dispersion forces, but it is held together by hydrogen bonding, a stronger force.

EXERCISE
9·11

Arrange the following compounds in order of increasing boiling point. Identify the predominant IMF in a liquid sample of each of the compounds, and explain your reasoning.

1. $H_2NCH_2CH_2OCH_3$ _____

2. $H_2NCH_2CH_2CH_2OH$ _____

3. $(CH_3)_2NCH_2OCH_3$ _____

pH and Amines

An amine can serve as a base in an acid-base reaction. Bases are proton acceptors (remember a proton is an H^+). In each example below, an acid donates an H^+ to the amine.

1. $CH_3 - \overset{\displaystyle H}{\underset{\displaystyle H}{N}} - H + HCl \rightarrow CH_3 - \overset{\displaystyle H}{\underset{\displaystyle H}{\overset{\oplus}{N}}} - H + \overset{..}{\underset{..}{:Cl}}{}^{\ominus}$

2. $CH_3 - \overset{\displaystyle H}{\underset{\displaystyle H}{N}} - CH_3 + HBr \rightarrow CH_3 - \overset{\displaystyle H}{\underset{\displaystyle H}{\overset{\oplus}{N}}} - CH_3 + \overset{..}{\underset{..}{:Br}}{}^{\ominus}$

3. ⬡ $- \overset{}{\underset{\displaystyle H}{N}} - H + HCl \rightarrow$ ⬡ $- \overset{\displaystyle H}{\underset{\displaystyle H}{\overset{\oplus}{N}}} - H + \overset{..}{\underset{..}{:Cl}}{}^{\ominus}$

Complete the following acid-base reactions involving amines.

1. $CH_3CH_2 — \overset{\cdot\cdot}{N} — H + HCl \longrightarrow$ _____
   |
   H

2. $CH_3CH_2 — \overset{\cdot\cdot}{N} — CH_3 + CH_3 — \overset{\overset{:O:}{\|}}{C} \diagdown_{:O.—H} \longrightarrow$ _____
   |
   H

When placed in water, water soluble amines establish an equilibrium system. The amine functional group reacts with water to accept a proton from the water. This reaction generates small quantities of hydroxide ions. For instance, propylamine reacts with water to produce the *n*-proply ammonium ion and a hydroxide ion. The reaction is

$$CH_3CH_2CH_2NH_2 + H_2O \rightleftharpoons CH_3CH_2CH_2NH_3^+ + OH^- \qquad K_b = 3.9 \times 10^{-4}$$

The equilibrium constant is $K_b = 3.9 \times 10^{-4}$

$$K_b = \frac{[CH_3CH_2CH_2NH_3^+][OH^-]}{[CH_3CH_2CH_2NH_2]} = 3.9 \times 10^{-4}$$

Amines are considered bases because aqueous solutions have a pH greater than 7.0.

In a similar way to the carboxylic acids discussed in the previous chapter, organic amines are weakly basic, meaning that most of the molecules in an aqueous solution are not reacting. The few that are disturb the water equilibrium in an amount determined by their equilibrium constant.

Propylamine has a base ionization constant of 3.9×10^{-4}.

When there are more hydroxide ions than hydronium ions present in solution, the pH is above 7.0, and the solution is said to be alkaline or basic.

The reason why water-soluble amines are considered weak bases is that not all of the base molecules react with the water. In fact, very few do. In technical terms, most amines are weakly basic because in aqueous solutions, the amine stays mostly intact as molecules. The percent of molecules that actually undergo the reaction with water is called the percent dissociation. Water-soluble organic amines dissociate between 1% and 15% in aqueous solution.

This means that about 1 molecule of amine for every 100 molecules dissociates. In contrast, strong bases dissociate 100%.

Let's assume that an aqueous solution of propylamine is about 5% dissociated and that we start with 100 molecules. We can set up an ICE table, where I stands for "initial," C stands for "change," and E stands for "equilibrium," to help track what is happening to each substance.

$$CH_3CH_2CH_2NH_2 + H_2O \rightleftharpoons CH_3CH_2CH_2NH_3^+ + OH^-$$

Initial	100 molecules	0	0
Change	−5 molecules	+5 ions	+5 ions
Equilibrium	95 molecules	5 ions	5 ions

% dissociation = number of OH⁻ ions at equilibrium/initial molecules of acid × 100% = 5%

$$\% \text{ dissociation} = \frac{5}{100} \times 100 = 5\%$$

This indicates that the extent of dissociation depends on the initial concentration of the base. A solution of water and propylamine consists largely of water molecules, along with a few hydroxide ions, a few propylammonium ions, and lots of propylamine molecules.

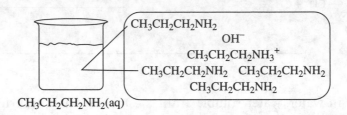

$CH_3CH_2CH_2NH_2(aq)$

Since the pH of a basic solution depends on the hydroxide ion concentration, the more $-OH^-$ ions in solution, the more basic the solution. It is possible to determine the pH of a solution of amine in water either by measuring the pH using a pH meter or by calculating the concentration of hydroxide ions. The initial concentration of the amine can't be used in the calculation because only a few of the molecules actually react with water to form hydroxide ions.

We can calculate the pH of a 0.100 M solution of proplyamine as follows.

$$CH_3CH_2CH_2NH_2 + H_2O \rightleftharpoons CH_3CH_2CH_2NH_3^+ + OH^- \quad K_b = 3.9 \times 10^{-4}$$

The ICE expression would be:

$$CH_3CH_2CH_2NH_2 + H_2O \rightleftharpoons CH_3CH_2CH_2NH_3^+ + OH^-$$

Initial	0.10 M	0	0
Change	$-x$	$+x$	$+x$
Equilibrium	0.10 $M - x$	x	x

$$K_b = \frac{(x)(x)}{0.10\ M - x} = 3.9 \times 10^{-4}$$

Assuming that x is small and we can ignore the $-x$ term, $x^2 = 3.9 \times 10^{-5}$,

$$x = 6.2 \times 10^{-3}\ M$$

This means that the concentrations at equilibrium are 0.0062 M OH⁻, 0.0062 M $CH_3CH_2CH_2NH_3^+$, and 0.10 $CH_3CH_2CH_2NH_2$. Calculating the concentration of the hydronium yields $K_w = [OH^-][H_3O^+] = 1.0 \times 10^{-14}$ $[H_3O^+] = 1.6 \times 10^{-12}\ M$, and substituting into the pH equation, we have pH = $-\log [1.6 \times 10^{-12}\ M] = 11.79$. Thus the result is a pH of 11.79, which is alkaline.

The magnitude of the base ionization constant indicates the extent to which the reaction favors the products, or, in other words, how alkaline is the solution formed. When two different bases of the same concentration are compared, the K_b values can indicate which of the two will be the stronger base. Keep in mind that this does not mean it is a strong base—just that it is stronger than another as more dissociation has occurred.

Table 9-2 Base Ionization Constants of Four Amines

	K_b
Ethylamine	5.6×10^{-4}
Methylamine	4.4×10^{-4}
Trimethylamine	6.4×10^{-5}
Aniline	3.9×10^{-10}

EXERCISE 9·13

Of the amines listed in Table 9-2, choose the base that generates the most hydroxide ions when placed in water. Write the reaction equation and the equilibrium expression.

Assume a 0.10 M concentration of the base chosen in Exercise 9-13, and solve for the concentration of the hydronium ion.

Solve for the pH of the base chosen in Exercise 9-13.

Of the bases listed in Table 9-2, choose the base that generates the fewest hydroxide ions when placed in water. Write the reaction equation and the equilibrium expression.

Assume a 0.10 M concentration of the base chosen in Exercise 9-16, and solve for the concentration of the hydroxide ions.

Solve for the pH of the base chosen in Exercise 9-17.

Calculate the percent dissociation for the strongest and weakest base. Do your answers for Exercises 9-13 and 9-16 align with your choice of the stronger base and the weaker base, respectively?

Amides

Amides have an amine group or a substituted amine group bonded to a carbonyl.

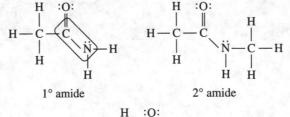

Just like amines, amides are classified as primary, secondary, or tertiary, depending on how many carbon groups are bonded to the nitrogen.

1° amide   2° amide

EXERCISE 9·20

Identify each of the following as a primary, secondary, or tertiary amide.

1. CH₃CH₂CH₂CH₂N̈—H
   |
   H

_____

2. CH₃—CHCH₂CH₃
        |
   H—N̈—CH₃

_____

3. CH₃CH₂CH₂CH₂—N̈—H
                |
                CH₃

_____

4.   CH₃CHCH₂CH₃
         |
   CH₃—N—CH₃

_____

5.   CH₃CHCH₂CH₃
         |
     H—N̈—H

_____

To name an amide, first identify the nearest structural carboxylic acid containing acyl group, next drop the suffix -ic of the acid, and add -amide. Identify the longest continuous carbon chain of the amide group. This will serve as the parent name. Use "N" to indicate any substitutions for other hydrogens on the nitrogen. For example, the following structure is named N-butylpropanamide.

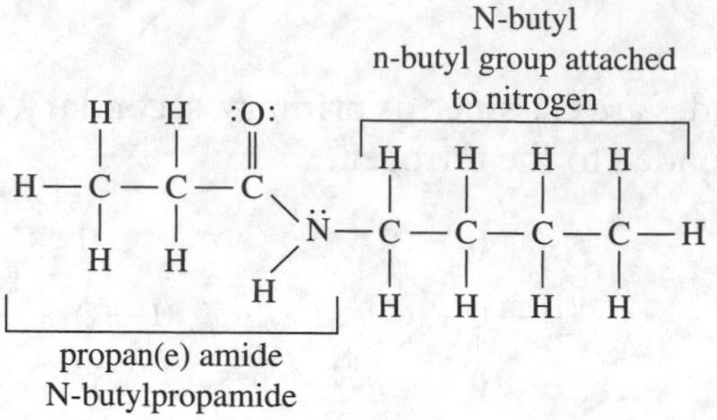

propan(e) amide
N-butylpropamide

EXERCISE
9·21

Name each of the amides in Exercise 9-20.

1. _____

2. _____

3. _____

4. _____

5. _____

Organic Reactions

This chapter consists of an introduction and a brief overview of how several chemical reactions that involve organic compounds occur. To understand organic reactions, it is crucial to recognize the types of organic compounds examined in the previous chapters.

Acid-Base Reactions

Proton, H⁺, exchange involving weak acids and water

Organic acids are weak acids that are called carboxylic acids. The simplest reaction that they undergo can be illustrated by placing an acid in water, where an exchange of a proton between two species occurs. A substance that donates a hydrogen ion is classified as an acid. A substance that accepts a hydrogen ion is classified as a base. Water is amphoteric; that is, it can serve as an acid or as a base, depending on what it is paired with. For example, acetic acid placed in water donates its acidic hydrogen ion, H^+, to water. Acetic acid serves as an acid and water serves as a base. In other words, acetic acid is a stronger acid than water. To fully visualize these reactions, it is important to be able to draw Lewis structures.

When acting as an acid, water donates an H^+ and forms an OH^- ion. When reacting with ammonia, water donates an H^+ to form the ammonium ion and hydroxide ion.

In a chemical equation, a double arrow between reactants (the starting species) and products (the ending species) indicates that the reaction goes to an equilibrium state. This means that as products are formed, they reverse their role, become reactants, and begin to make product. Such reversals continue, ensuring that, at equilibrium, the rate of making product equals the rate of making reactant. This does not mean that the amounts of product and reactant are equal. Reactions with weak acids go to equilibriums with very small amounts of product.

The two models that follow show, first via models and then by means of Lewis structures, the transfer of the H^+ to the water. Lewis structures enable you to see where to place the H^+ so that a pair of electrons is shared. Remember that an H^+ is a proton with no electrons or neutrons present.

Hydrolysis reactions (reactions with water)

$$CH_3 - \overset{\overset{\displaystyle :O:}{\|}}{C} - \overset{..}{\underset{..}{O}} - \textcircled{H} \; + \; H - \overset{..}{\underset{|}{O}}: \; \rightleftharpoons \; \left[H - \overset{\textcircled{H}}{\underset{|}{O}}: \right]^{+} \; + \; CH_3 - \overset{\overset{\displaystyle :O}{\|}}{C} - \overset{..}{\underset{..}{O}}:^{-}$$

acetic acid

The three hydrogen atoms bonded directly to a carbon atom in acetic acid are not involved in the chemical reaction because they are too strongly bonded to be donated as "acidic hydrogen atoms." Only the hydrogen attached to the oxygen is considered acidic. Remember that the functional group containing an acid is the COOH group. It takes a bit more energy to break a C-H bond than to break the O-H bond in the carboxylic acid function group COOH.

acidic hydrogen

$$H - \overset{\overset{\displaystyle H}{|}}{\underset{\underset{\displaystyle H}{|}}{C}} - \overset{\overset{\displaystyle :O:}{\|}}{C} - \overset{..}{\underset{..}{O}} - H$$

One of the driving forces of this reaction is the formation of the acetate ion, which has a resonance (has more than one possible structure) for its stabilized anion. The electron density between one oxygen atom, the carbon atom bonded to the oxygen atom, and the other oxygen atom is delocalized. This imparts stability to the anion. Note that ion Lewis structures are drawn inside a bracket with the charge on the outside as a superscript.

$$\left[H - \overset{\overset{\displaystyle H}{|}}{\underset{\underset{\displaystyle H}{|}}{C}} - C \overset{\displaystyle :O:}{\underset{\displaystyle :O:}{}} \right]^{-} \; \rightleftharpoons \; \left[H - \overset{\overset{\displaystyle H}{|}}{\underset{\underset{\displaystyle H}{|}}{C}} - C \overset{\displaystyle :\overset{..}{O}:}{\underset{\displaystyle :O:}{}} \right]^{-}$$

Because of the equilibrium, most of the acetic acid molecules remain as molecules, and only a few hydrogen atoms on the acetic acid react with water molecules to form some hydrogen ions and the acetate ion. About 4% of the acetic acid molecules react with water. The number of molecules reacting per 100 molecules is known as the percent ionization. A low percent ionization is a characteristic of weak acids. Acetic acid, along with many more organic carboxylic acids, is classified as a weak acid. Note that the hydronium ion, H_3O^+, is a stronger acid than the acetate ion, so the reverse reaction will occur.

$$CH_3COO^-(aq) + H_3O^+(aq) \rightleftharpoons CH_3COOH^-(aq) + H_2O(l)$$

Reading the reaction from left to right, one can classify each reactant as either an acid or a base, and one can classify each product as either the conjugate acid or the conjugate base. Thus in the reverse reaction, the conjugatesare the acid and the base.

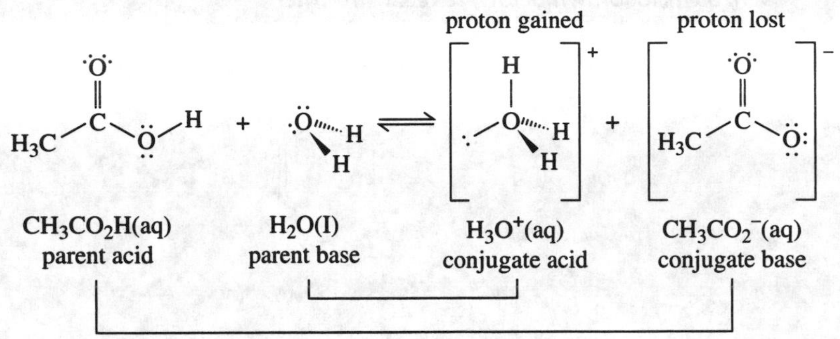
proton gained
proton lost
$CH_3CO_2H(aq)$
parent acid
$H_2O(l)$
parent base
$H_3O^+(aq)$
conjugate acid
$CH_3CO_2^-(aq)$
conjugate base

EXERCISE

10·1

For each of the following acids, draw the Lewis structure and circle the acidic hydrogen.

1. propanoic acid

2. methanoic acid

3. ethanoic acid

EXERCISE

10·2

Write the balanced chemical equation for the equilibrium system that is established when benzoic acid (drawn below) reacts with water.

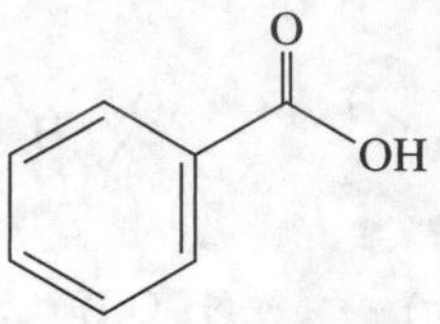

EXERCISE

10·3

Draw the balanced chemical equation for the equilibrium reaction of methanoic acid with water. Indicate the acid and base and the conjugate acid and conjugate base in the reaction.

Draw the balanced chemical equation for the equilibrium reaction of ethanoic acid in water. Indicate the acid and base and the conjugate acid and conjugate base in the reaction.

Explain the concept of percent ionization. To what degree (percentage) do weak acids ionize? What species is the predominant species in any weak acid organic reaction in aqueous solution?

Weak Acid—Strong Base Reactions

As you learned in the previous section, and in Chapter 8, the carboxylic acid functional group has an acidic hydrogen. This hydrogen ion will react with strong bases, such as sodium hydroxide or potassium hydroxide, to produce a salt and water. Salts are made from the conjugate base and conjugate acid ions. Salts can be acidic, basic, or neutral, depending on the ions making up the salt.

Propanoic acid reacts with the hydroxide ion to form the propanoate ion and water:

$$CH_3CH_2COOH(aq) + K^+(aq) + OH^-(aq) \rightarrow CH_3CH_2COO^-(aq) + K^+(aq) + H_2O(l)$$

Because potassium hydroxide is an ionic compound, it dissolves to form potassium ions, K^+, and hydroxide ions, OH^-, in water and is written as ions in an ionic equation. The OH^- ions react with the acidic hydrogen on the propanoic acid to form water. What remains are propanoate ions and potassium ions. The spectator ion, K^+, is removed from a net ionic equation.

$$CH_3CH_2COOH(aq) + OH^-(aq) \rightarrow CH_3CH_2COO^-(aq) + H_2O(l)$$

      acid   base   conjugate base   conjugate acid

EXERCISE
10·6

Write equations for the acid-base ionic reaction and the net ionic equation for the reaction between benzoic acid and sodium hydroxide. Indicate the acid and base conjugate pairs.

EXERCISE 10·7

Draw the reaction in Exercise 10-6, and show the hydrogen transfer.

EXERCISE 10·8

Write equations for the acid-base ionic reaction and the net ionic equation for the reaction between acetic acid and sodium hydroxide.

Weak Acid—Weak Base Reactions

Weak acids can react with weak bases. For example, ammonia can react with water as follows:

$$\underset{\substack{\text{hydrogen} \\ \text{ion acceptor:} \\ \text{B-L base}}}{\text{H}-\overset{\text{H}}{\underset{\text{H}}{\text{N}}}:} + \underset{\substack{\text{hydrogen} \\ \text{ion donor:} \\ \text{B-L acid}}}{\text{H}-\overset{}{\underset{\text{H}}{\text{O}}}:} \rightarrow \text{H}-\overset{\text{H}}{\underset{\text{H}}{\text{N}^+}}-\text{H} + :\overset{}{\underset{\text{H}}{\text{O}}}:^-$$

In this reaction, water serves as the weak acid (a proton donor), and ammonia serves as a weak base (a proton acceptor). Note that one of the two lone-electron pairs on nitrogen serves as the two electrons in the newly formed covalent bond between nitrogen and hydrogen. The two electrons on oxygen in the water remain with the oxygen to form a lone pair on the hydroxide ion.

Weak acids donate their hydrogen to the nitrogen in a weak base such as ammonia to form the ammonium ion.

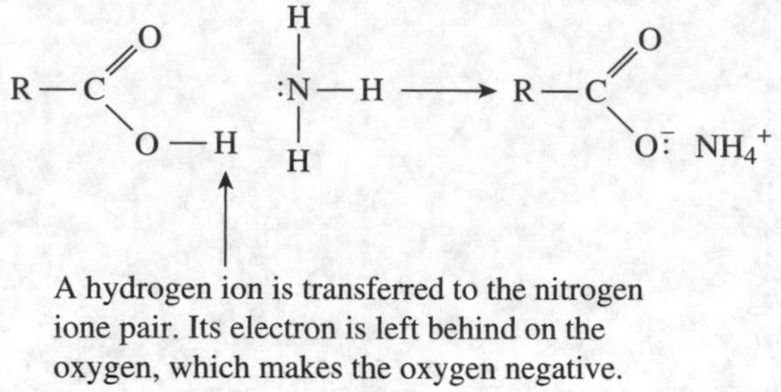

$$R-C \overset{O}{\underset{O-H}{}} \quad \overset{H}{\underset{H}{:N-H}} \longrightarrow R-C \overset{O}{\underset{O^-}{}} NH_4^+$$

A hydrogen ion is transferred to the nitrogen
ione pair. Its electron is left behind on the
oxygen, which makes the oxygen negative.

EXERCISE

10·9

Write and draw the chemical reaction between propanoic acid and methylamine.

EXERCISE

10·10

Complete and balance each of the following chemical equations.

1. $CH_3CH_2CH_2COOH + CH_3CH_2NH_2 \rightleftharpoons$

2. $C_6H_5COOH + CH_3CH_2NH_2 \rightleftharpoons$

3. $HCOOH + (CH_3CH_2)_2NH \rightleftharpoons$

4. $C_6H_5COOH + C_6H_5CH_2NH_2 \rightleftharpoons$

For each reaction in Exercise 10-10, draw the reaction and indicate the proton transfer.

1.

2.

3.

4.

Reactions to Form Carboxylic Acids

There are three basic reactions that form organic acids. These different reactions involve organic compounds such as acyl chlorides, alcohols, aldehydes, and acid anhydrides.

An acyl group has a formula of RCO. The C=O group is a carbonyl group. Acyl groups are the basis of esters, aldehydes, ketones, anhydrides, amides, acid chlorides, and carboxylic acids.

- Acetyl chloride reactions with water produce carboxylic acids.

When the R group in the acyl structure is a methyl group, it is called the acetyl group and it has the chemical structure CH_3CO. The other bond attached to the carbon can be a —OH, —NH$_2$, —X, —R, or —H.

Acetyl chlorides have the following structure:

$$R - \overset{\overset{\displaystyle :O:}{\|}}{C} - \ddot{\underset{..}{Cl}}:$$

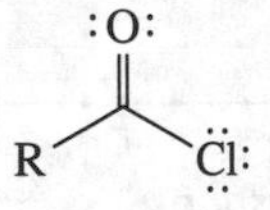

Low-molecular-weight acetyl chlorides react with water to produce carboxylic acids according to the following mechanism. (Low-molar mass or molecular-weight acetyl chlorides are considered to be those that are below 100 g/mol and contain fewer than six carbon atoms.)

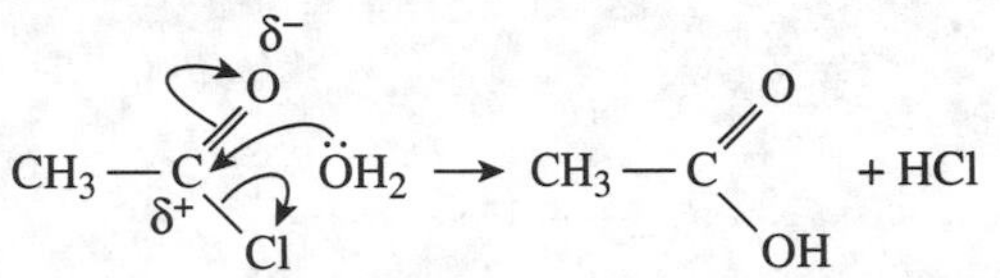

$$CH_3 - \overset{\overset{\displaystyle \overset{\delta-}{O}}{\|}}{\underset{\underset{\displaystyle Cl}{|}}{\underset{\delta+}{C}}} \quad \ddot{O}H_2 \longrightarrow CH_3 - \overset{\overset{\displaystyle O}{\|}}{C} - OH \quad + HCl$$

Acetyl chloride reacts with water to produce acetic acid and hydrogen chloride.

- Reactions of acid anhydrides with water produce carboxylic acids.

Acid anhydrides have the following structure:

$$R - \overset{\overset{\displaystyle \cdot\ddot{O}\cdot}{\|}}{C} - \underset{\cdot\ddot{O}\cdot}{} - \overset{\overset{\displaystyle \cdot\ddot{O}\cdot}{\|}}{C} - R$$

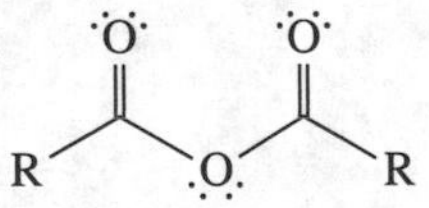

Acid anhydrides can be recognized by double carbonyl groups bonded together by a bridging oxygen atom.

Low-molar mass or molecular-weight acid anhydrides react with water to produce carboxylic acids. Low-mass acid anhydrides are considered to be those that are below 120 g/mol and contain fewer than eight carbon atoms.

$$CH_3\overset{\overset{\displaystyle O}{\|}}{C} - O - \overset{\overset{\displaystyle O}{\|}}{C} - CH_3 + H_2O \longrightarrow 2\, CH_3\overset{\overset{\displaystyle O}{\|}}{C} - OH$$

Acetic anhydride reacts with water to produce acetic acid.

$$CH_3\overset{\overset{\displaystyle O}{\|}}{C} - O - \overset{\overset{\displaystyle O}{\|}}{C} - CH_3 + H_2O \longrightarrow 2\, CH_3\overset{\overset{\displaystyle O}{\|}}{C} - OH$$

♦ Oxidation of primary alcohols or aldehydes produces carboxylic acids.

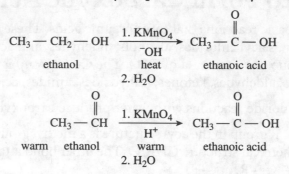

Answer the following questions.

1. What is the functional group in organic acids? _____

2. Draw a chemical equation to illustrate how a primary alcohol oxidizes to form a carboxylic acid.

_____

3. Draw a chemical equation to illustrate how propanol in the presence of a catalyst can form a carboxylic acid.

_____

4. Methanol in the presence of CrO_3 and H^+ yields what carboxylic acid? _____

5. Give an example of an aldehyde that can form a carboxylic acid. Draw the reaction.

_____

Formation of Esters

Carboxylic acids react with alcohols in the presence of a small amount of acid (acting as a catalyst) to produce an ester. Acetic acid reacts with ethanol to produce the ester ethyl acetate and water. Note that in the equation, the catalyst is indicated over the arrow because it is neither a reactant nor a product in the overall reaction.

$$CH_3-C \overset{:O:}{\underset{:O-H}{}} \quad + HOCH_2CH_3 \xrightarrow{H^+} \quad CH_3-C \overset{:O:}{\underset{:O-CH_2CH_3}{}} \quad + H_2O$$

acetic acid   ethanol   ethyl acetate

EXERCISE
10·13

Write the products and balance the chemical equation for the formation of an ester when propanoic acid reacts with methanol in the presence of some acid.

$$CH_3CH_2COOH + CH_3OH \underset{\longleftarrow}{\xrightarrow{H^+}} \underline{\hspace{4cm}}$$

When ethanoic acid reacts with propanol in the presence of some acid, propyl ethanoate forms.

$$CH_3COOH + CH_3CH_2CH_2OH \overset{H^+}{\rightleftharpoons} CH_3CH_2CH_2OOCH_3 + H_2O$$

ethanoic acid   propanol   water   propyl-ethanoate

In this reaction, the ester forms when the OH group from the acid reacts with the hydrogen from the alcohol.

EXERCISE

10·14

Show what you would do to produce each of the following esters. In other words, write the balanced chemical equation for the formation of each ester.

1. ethyl propanoate

2. propyl methanoate

3. methyl butanoate

4. propyl ethanoate

Esters often have pleasant odors and different esters are responsible for the sweet odor of fruits.

$$CH_3CH_2-C\overset{\displaystyle O}{\underset{\displaystyle O-CH_2-CH_3}{\big<}}$$

ethyl propanoate

$$H-C\overset{\displaystyle O}{\underset{\displaystyle O-CH_2-CH_2-CH_3}{\big<}}$$

propyl methanoate

$$CH_3CH_2CH_2-C\overset{\displaystyle O}{\underset{\displaystyle O-CH_3}{\big<}}$$

methyl butanoate

$$CH_3-C\overset{\displaystyle O}{\underset{\displaystyle O-CH_2-CH_2-CH_3}{\big<}}$$

propyl ethanoate

Anhydrides react with phenols to produce esters and a carboxylic acid.

$$CH_3-C\overset{O}{\big<} + H-\ddot{N}-H \longrightarrow CH_3-C\overset{O}{\big<} + CH_3COOH$$

Hydrolysis of Esters

One way to synthesize aspirin is to react salicylic acid with acetic anhydride. Aspirin is an example of an ester.

salicylic acid   acetic anhydride   aspirin   acetic acid

Esters react with water and a small amount of acid to produce the original carboxylic acid and the original alcohol from which it was made.

$$R-\overset{\displaystyle O}{\overset{\displaystyle \|}{C}}-OR + H_2O \underset{}{\overset{H^+}{\rightleftharpoons}} R-\overset{\displaystyle O}{\overset{\displaystyle \|}{C}}-OH + ROH$$

an ester   water   a carboxylic   an alcohol
                                  acid

When heated, esters react with water to produce the original carboxylic acid and the original alcohol from which it was made.

$$R-\overset{\displaystyle O}{\overset{\displaystyle \|}{C}}\diagdown_{OR'} \underset{heat}{\overset{H_2O}{\longrightarrow}} R-\overset{\displaystyle O}{\overset{\displaystyle \|}{C}}\diagdown_{OH} + R'OH$$

The mechanism for this reaction starts with the reaction of an H⁺ ion from the catalyst reacting with the ester. Water reacts to form a complex ion. A hydrogen shifts over to the nearby oxygen group. Next alcohol is eliminated. Then a hydrogen ion withdraws, leaving the carboxylic acid. Many organic reactions take place in a series of steps like this one does.

For example, butyl acetate reacts with water and a small amount of acid to produce acetic acid and 1-butanol (n-butyl alcohol).

$$CH_3COCH_2CH_3CH_3 + H_2O \xrightleftharpoons{H^+} CH_3-\overset{O}{\overset{\|}{C}}-OH + CH_3CH_2CH_2CH_2OH$$

| butyl acetate | water | acetic acid | 1-butanol |
| | | | (butyl alcohol) |

Write the chemical equation for the hydrolysis of each of the following esters.

1. $CH_3-\overset{O}{\overset{\|}{C}}-O-CH_2CH_2CH_2CH_3 + H_2O \xrightleftharpoons{H^+}$
   butyl acetate

2. $CH_3CH_2CH_2\overset{O}{\overset{\|}{C}}-O-CH_2CH_3 + H_2O \xrightleftharpoons{H^+}$
   ethyl butanoate

3.

   acetyl
   salicylic acid

Reaction of Amides

Amides can be made by reacting an ester and an amine. For example, propanoic acid reacts with methylamine to form N-methylpropanamide and water.

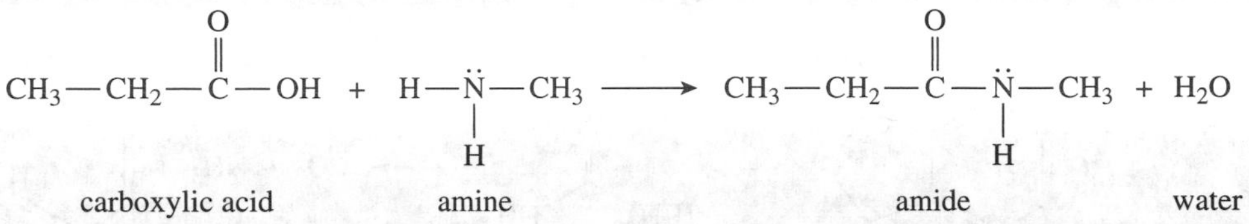

$$CH_3-CH_2-\overset{\overset{\displaystyle O}{\|}}{C}-OH \ + \ H-\overset{\overset{\displaystyle \cdot\cdot}{N}}{\underset{\underset{\displaystyle H}{|}}{}}-CH_3 \ \longrightarrow \ CH_3-CH_2-\overset{\overset{\displaystyle O}{\|}}{C}-\overset{\overset{\displaystyle \cdot\cdot}{N}}{\underset{\underset{\displaystyle H}{|}}{}}-CH_3 \ + \ H_2O$$

carboxylic acid   amine   amide   water

The following reaction illustrates the general mechanism for the reaction of an ester with ammonia. In the first step the partial negative charge of the nitrogen in ammonia uses the two lone pair electrons to attack the partial positive carbon atom in the carbonyl group of the ester. A H$^+$ ion is relocated to the oxygen. The two electrons on the carbonyl oxygen shift to create a double bond between oxygen and carbon. This forces the two electrons in the bond between carbon and oxygen to break the temporary C-O bond and from an alcohol, H-O-R. In this process an amide is created.

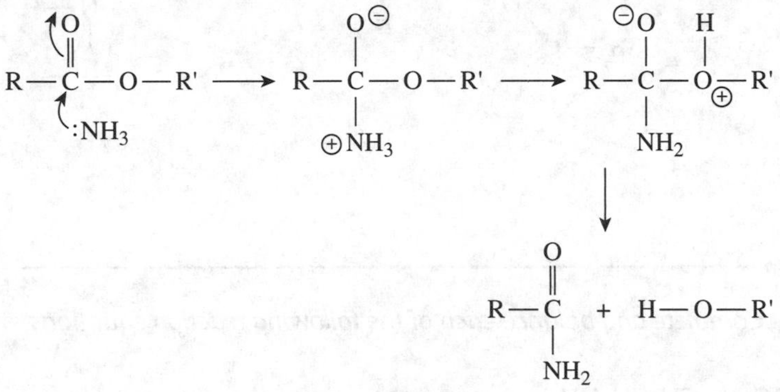

The following reaction illustrates the general mechanism for the reaction of an ester with n-propylamine. In the first step the partial negative charge of the nitrogen in propylamine uses the two lone pair electrons to attack the partial positive carbon atom in the carbonyl group of the ester. A H$^+$ ion is relocated to the oxygen. The two electrons on the carbonyl oxygen shift to create a double bond between oxygen and carbon. This forces the two electrons in the bond between carbon and oxygen to break the temporary C-O bond and from an alcohol, H-O-R. In this process an amide is created.

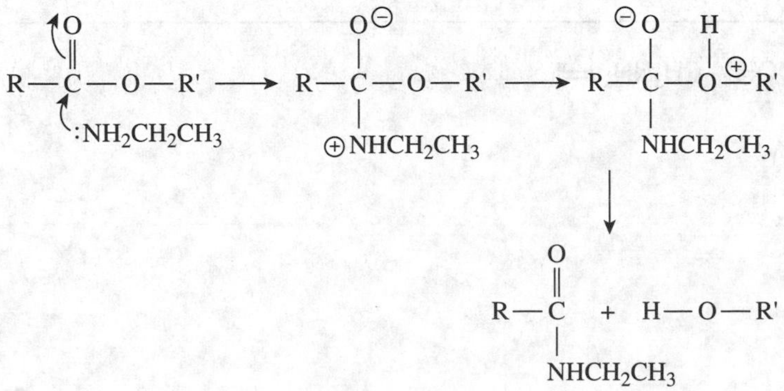

Amides and carboxylic acids can be made by reacting anhydrides with amines. For example, phthalic anhydride reacts with diaminomethanal to produce carbon dioxide, ammonia, and phthalimide (IUPAC name Isoindole-1,3-dione). Because phthalimide has a "hidden" N-H group, it serves as a weak base, similar to ammonia. Phthalimide is used as a starting point to make other compounds.

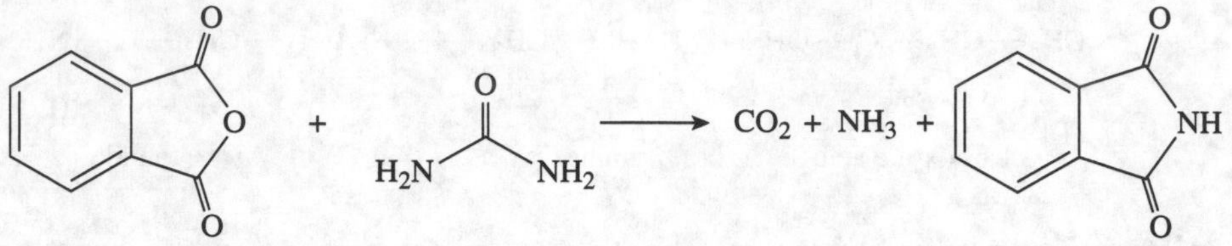

Acetic anhydride reacts with ammonia to form acetamide (N-methylformamide) and acetic acid. Acetamide can be used as a starting reagent to make plastics. Acetamide is also used as an industrial solvent, when a weak base is required.

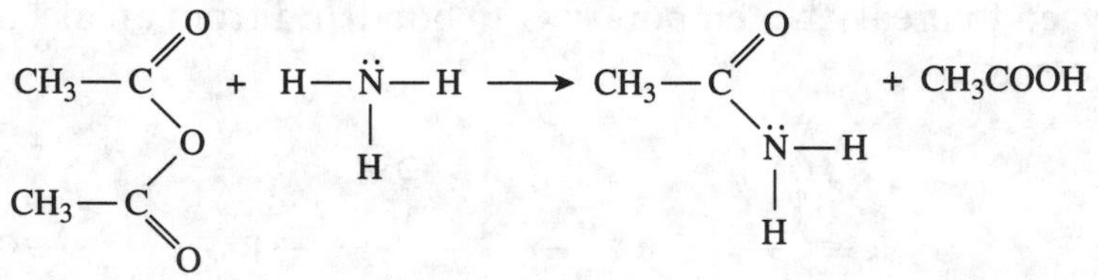

EXERCISE
10·16

Complete and balance each of the following reaction equations.

1. $CH_3CH_2COOH + CH_3CH_2NH_2 \rightleftharpoons$

2. $CH_3CH_2COOH + CH_3NH_2 \rightleftharpoons$

3. $CH_3CH_2COOH(aq) + K^+(aq) + OH^-(aq) \rightleftharpoons$

4. $CH_3CH_2COOH + CH_3OH \rightleftharpoons$

5. $CH_3CH_2CH_2CH_2-\overset{\cdot\cdot}{\underset{\cdot\cdot}{O}}-\overset{\overset{\displaystyle :O:}{\|}}{C}-CH_3 + H_2O \rightleftharpoons$

Proteins are one of the main components of all living matter. Proteins are involved in the processes of growth and reproduction. Our study of amides provides a basis for understanding amino acids, peptides, and proteins. A protonated form of (+)-alanine reveals it has two function groups, a carboxylic acid group and a protonated amine group.

a protonated α (alpha)-aminocarboxylic acid
an amino acid

An amino acid is a carboxylic acid with a protonated amino group on the α (alpha) carbon. The general form of an amino acid is

$$
\begin{array}{c}
:O: \\
\| \\
R \cdots C \cdots H \quad C - \ddot{O} - H \\
| \\
H - N - H \\
| \oplus \\
H
\end{array}
$$

Protein molecules are long chains of amino acids linked together by amide bonds. An example of a general form for a dipeptide is

a dipeptide

$$
\begin{array}{c}
:O: \qquad\qquad H \quad :O: \\
\| \qquad\qquad | \quad \| \\
R \cdots C \cdots C \quad N - C - C \quad N - \\
\quad H \quad | \quad R \quad | \\
\quad H \qquad\qquad H
\end{array}
$$

amide bonds or
peptide linkages

Amino acids are linked together by amide bonds also known as peptide linkages. The link is between the carboxylic acid groups of one amino acid and the amino group in a second amino acid.

There are about 20 amino acids of importance to building proteins. These amino acids can be combined in many different sequences. Some amino acids are positively charged, some have no charge, and some have a negative charge.

$$
\begin{array}{c}
:O: \\
CH_3 \quad \| \\
| \\
H - C - C - \ddot{O}:^{\ominus} \\
| \\
CH_3 \quad NH_3 \oplus
\end{array}
\qquad
\begin{array}{c}
:O: \\
CH_3 \quad \| \\
| \\
H - C - CH_2 \quad C - \ddot{O}:^{\ominus} \\
| \\
CH_3
\end{array}
$$

(+) – valine   (+) – lencine
val   len

$$
\begin{array}{c}
:O: \\
CH_3 \quad \| \\
| \\
H - C - C - \ddot{O}:^{\ominus} \\
\| \\
CH_2 \\
CH_3
\end{array}
\qquad
\begin{array}{c}
:O: \\
\| \\
C - \ddot{O}:^{\ominus} \\
C \cdots H \\
CH_2 \quad N \cdots H \\
| \quad \oplus \quad H \\
CH_2 \\
\backslash CH_2
\end{array}
$$

(+) – isolencine   (–) proline
ile   pro

When 50 or more amino acids are linked together, the structure is called a polypeptide. A protein might consist of one long polypeptide chain or it might consist of several long chains. The study of amino acids, peptides, and proteins is fascinating. If you are interested, you should take a biochemistry course after mastering organic chemistry.

Reactions of Aldehydes

The main structural feature of aldehydes is the carbonyl group. Oxygen, with an electronegativity of 3.0, is more electronegative than carbon, with an electronegativity of 2.5. Remember that the shared electrons in the bond will move more toward the more electronegative atom. In the carbon-oxygen double bond, oxygen attracts the electrons in the bond, resulting in an unequal sharing of the electrons, and this creates a polar covalent bond. The oxygen in the carbonyl has a partial negative charge, and the carbon has a partial positive charge.

$$\delta^- \longleftarrow \text{attracted to } \delta^+ \text{ electrophiles}$$
$$\text{O}$$
$$-\text{C}$$
$$\delta^+ \qquad \text{H}$$
$$\uparrow$$
$$\text{attracted to } \delta^- \text{ nucleophile}$$

Nucleophiles have a negative charge and will attack the positive-charged carbon in a carbonyl group. Electrophiles have a positive charge and will attack the negative-charged oxygen in a carbonyl group.

> As the complexity of the group attached to an aldehyde functional group increases, the reactivity of the carbonyl group decreases. For example, formaldehyde is more reactive to nucleophilic attack than is butanal.

EXERCISE 10·17

Rank the following compounds from greatest to least reactivity to nucleophilic attack.

_____ > _____ > _____ > _____

1. $CH_3CH_2CH{=}O$ _____

2. $CH_2{=}O$ _____

3. $CH_3CH_2CH_3CH{=}O$ _____

4. $CH_3CH_2CH_2CH_3CH{=}O$ _____

The formation of alcohols is due to the reaction of nucleophilies on the carbonyl carbon of aldehydes.

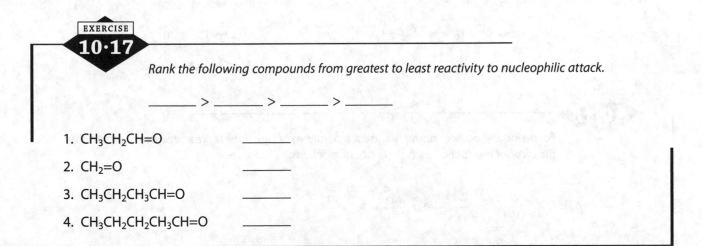

$$R{-}CH{=}O + :Nu^- \xrightarrow{H_2O} R{-}\underset{\underset{Nu}{|}}{\overset{\overset{OH}{|}}{C}}{-}H$$

The hydride ion H:⁻ is a good nucleophile. Sodium borohydride, $NaBH_4$, and lithium aluminum hydride, $LiAlH_4$, are both reagent sources containing the hydride ion. Treatment of an aldehyde with $NaBH_4$, followed by the addition of water, adds an H to the carbonyl oxygen and an H to the carbonyl carbon.

$$CH_3CH_2CH=O + H:^- \xrightarrow{H_2O} \begin{array}{c} OH \\ | \\ CH_3CH_2-C-H \\ | \\ H \end{array}$$

This reaction occurs in two steps:

$$\underset{\underset{BH_3}{\overset{|}{\underset{H}{|}}}{Na^+,}}{\overset{:O:}{\overset{\|}{CH_3CH_2\overset{}{C}-H}}} \longrightarrow \underset{\underset{BH_3}{\overset{|}{\underset{H}{|}}}}{\overset{:\ddot{O}:^-}{CH_3CH_2\overset{}{C}-H}} \overset{H-O-H}{\longrightarrow} \underset{\overset{|}{H}}{\overset{O-H}{CH_3CH_2\overset{}{C}-H}} + Na^+ OH^-$$

In this example, a hydride ion from the sodium borohydride first attacks the partial-positive-charged carbon in the propanal, leaving the carbonyl oxygen with a negative charge and the product BH_3. Then the negative-charged carbonyl oxygen is attracted to the positive hydrogen of water, forming the alcohol propanol and the base sodium hydroxide.

Nucleophiles attack here. $\overset{\delta^+}{\underset{}{C}}=\overset{\delta^-}{\ddot{O}}$

$$Nu:^- \quad \overset{}{\underset{}{C}}=\ddot{O} \rightleftharpoons Nu-\overset{|}{\underset{|}{C}}-\ddot{O}^- \xrightarrow{H_2O} \overset{Nu}{\underset{}{C}}=\ddot{O}-H + :\ddot{O}H^-$$

Following the pattern above, write each sequence of reactions represented below, and name the alcohol product of each sequence of reactions.

1. $H-C\equiv N: +$ (benzaldehyde structure) $\overset{:O:}{\overset{\|}{C}-H}$ $\xrightarrow{NaOH}$ A $\xrightarrow[heat]{H_3O^+}$ B

_____

_____

_____

_____

_____

_____

_____

$$2. \quad CH_3NH_2 \quad + \quad \overset{:O:}{\underset{\|}{CH_3C}} - H \quad \overset{H_2O}{\rightleftharpoons}$$

3. H_2NNH_2 + $CH_3CH_2\overset{\displaystyle :O:}{\underset{\displaystyle \|}{C}}$—H $\xrightarrow{\text{KOH}}$

4. CH_3OH + $CH_3CH_2\overset{\displaystyle :\!O\!:}{\overset{\|}{C}}$—$H$ $\underset{\text{catalyst}}{\overset{\text{HCl}}{\rightleftharpoons}}$

5. NaBH$_4$ + (benzaldehyde) $\xrightarrow{}$ A $\xrightarrow{H_3O^+}$ B

_____

_____

_____

_____

_____

_____

_____

_____

_____

_____

_____

_____

_____

_____

_____

_____

_____

_____

Reactions of Carboxylic Acids with Base

Carboxylic acids react with strong bases to produce a salt and water. The salts formed then undergo a hydrolysis reaction resulting in a basic solution.

For instance, aqueous acetic acid combined with aqueous sodium hydroxide forms the salt sodium acetate and water. The ionic salt sodium acetate is soluble, so it is actually a solution of acetate ions and sodium ions.

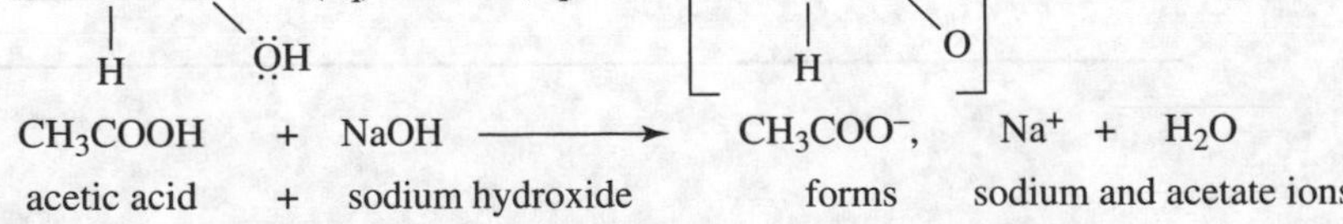

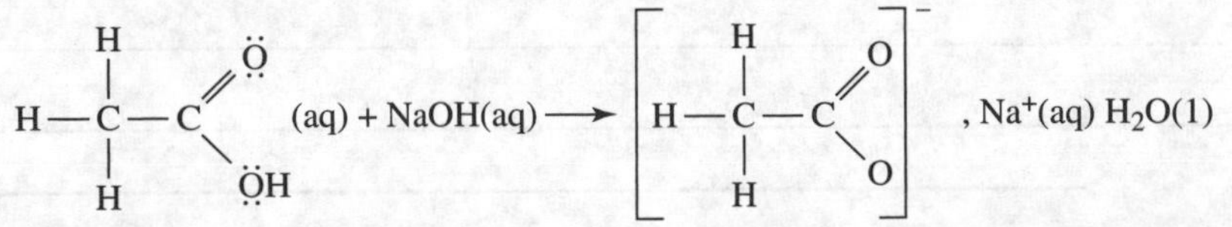

$$CH_3COOH + NaOH \longrightarrow CH_3COO^-, \quad Na^+ + H_2O$$

acetic acid + sodium hydroxide   forms   sodium and acetate ions and water

The acetate ion in the solution then undergoes a secondary reaction with the water that is called hydrolysis. The products of this reaction cause the resulting solution to be basic as a consequence of the formation of hydroxide ion.

$$CH_3COO^- + H_2O \rightleftharpoons CH_3COOH + OH^-$$

acetate ion   water   acetic acid + hydroxide ion

We can still determine the pH, but because hydroxide is present, the concentration of hydronium ions needs to be calculated first using the equation for the equilibrium constant of water itself. This relation is always obeyed in aqueous (water) solutions and shows that the concentration of hydroxide ions and hydronium ions are intimately related.

$$K_w = [H_3O^+][OH^-]$$

where K_w is the equilibrium constant of water, and $K_w = 1.0 \times 10^{-14}$. If the amount of hydroxide ion concentration is $0.001\ M$, then the concentration of hydronium is $1.0 \times 10^{-11}\ M$. This equates to a pH of 11.

EXERCISE

10·19

Complete and balance each of the following chemical equations, assuming a complete acid-base reaction. Then write the balanced equation for the ion from the salt, reacting with water to reach an equilibrium.

1. $CH_3CH_2COOH(aq) + KOH(aq)$

_____

_____

2. $HOOCCH_2CH_2COOH + LiOH(aq)$

_____

_____

3. $C_6H_5COOH + NaOH(aq)$

_____

_____

4. $HOOCC_6H_4COOH + LiOH(aq)$

_____

_____

EXERCISE

10·20

Explain why the solution formed by a weak acid reacting with a strong base is a basic solution. Include an equation in your explanation.

EXERCISE

10·21

Explain why organic acids are considered weak.

Answer Key

1   Carbon and the Study of Organic Chemistry

1-1   1. $14\,p^+$ 2. $8\,p^+$ 3. $15\,p^+$ 4. $17\,p^+$ 5. $7\,p^+$

1-2   1. $15\,n^0$ 2. $62\,n^0$ 3. $8\,n^0$ 4. $9\,n^0$ 5. $22\,n^0$

1-3   1. $^{39}_{19}K$ 2. $^{28}_{14}Si$ 3. $^{30}_{14}Si$ 4. $^{35}_{17}Cl$ 5. $^{11}_{5}B$

1-4   1. 13 protons and 11 neutrons   2. 27 protons and 33 neutrons
        3. 9 protons and 10 neutrons   4. 6 protons and 7 neutrons   5. 7 protons and 8 neutrons

1-5   1. 24.30 amu   2. 107.9 amu   3. 28.09 amu   4. 10.81 amu   5. 20.18 amu

1-6   1. $16\,e^-$ 2. $8\,e^-$ 3. $7\,e^-$ 4. $17\,e^-$ 5. $1\,e^-$

1-7   1. 6 protons and 10 electrons   2. 7 protons and 10 electrons
        3. 11 protons and 10 electrons   4. 20 protons and 18 electrons
        5. 16 protons and 18 electrons

1-8   1. P^{3-} 2. Al^{3+} 3. Zn^{2+} 4. Ne   5. F^-

1-9   1. 4 valence electrons   2. 5 valence electrons   3. 6 valence electrons
        4. 7 valence electrons   5. 6 valence electrons

1-10   1. 58.12 g/mol   2. 74.12 g/mol   3. 31.06 g/mol   4. 32.04 g/mol   5. 46.07 g/mol
        6. 60.09 g/mol   7. 112.98 g/mol   8. 26.04 g/mol   9. 180.16 g/mol   10. 46.07 g/mol

1-11

1-12   1. polar covalent   2. polar covalent   3. nonpolar covalent   4. polar covalent
        5. polar covalent

1-13   1. H_2O $H-\overset{..}{\underset{..}{O}}-H$ H—O bond is polar (1.4 EN difference); the molecule is polar since the vector dipoles do not cancel each other.

   $H \rightarrow O \leftarrow H$

   2. NH_3 $H-\overset{..}{\underset{|}{N}}\cdots H$ $H \rightarrow N \leftarrow H$ N—H bond is polar (0.9 EN difference); the molecule is polar since the vector dipoles do not cancel.
             H H

   3. $H-\overset{\overset{H}{|}}{\underset{\underset{H}{|}}{C}}-\overset{\overset{H}{|}}{\underset{\underset{H}{|}}{C}}-H$ C—H bond is nonpolar (0.4 EN difference); the molecule is nonpolar since all vectors cancel.

   4. $H-\overset{\overset{H}{|}}{\underset{\underset{H}{|}}{C}}-\overset{\overset{:O:}{||}}{C}$:O:—H   C—H bond is nonpolar (0.4 EN difference), C—C bond is nonpolar (0 EN difference), H—O bond is polar (1.4 EN difference), and C—O bond is polar (0.9 EN difference); the molecule is polar since not all vectors cancel and the molecule is not symmetrical.

   $\overset{O}{\underset{O}{C}}\rightarrow H$

   5. $H-\overset{\overset{H}{|}}{\underset{\underset{H}{|}}{C}}\overset{\overset{\overset{\delta-}{:O:}}{||}}{\underset{\delta+}{C}}\overset{\overset{H}{}}{\underset{\underset{H}{|}}{C}}-H$ $\overset{\delta-}{O}\uparrow\underset{C}{\delta+}$ C—H bond is nonpolar (0.4 EN difference), C—C bond is nonpolar (0 EN difference), and C—O bond is polar (0.9 EN difference); the molecule is polar since not all vectors cancel and the molecule is not symmetrical.

1-14

MOLECULE	DISPERSION FORCE	DIPOLE-DIPOLE FORCE	HYDROGEN BOND
1. H_2O	X	X	X
2. NH_3	X	X	X
3. C_2H_6	X		
4. $C_2H_4O_2$	X	X	X
5. C_3H_6O	X	X	

1-15   The structure gives the relative positions of the atoms and the connectivity of the atoms. A 3-D drawing allows vector dipoles to be determined. Whether or not dipoles cancel determines whether the molecule is polar or nonpolar, therefore the structure is more helpful than the formula.

2   Alkanes

2-1   1. but- = 4 C in the chain, for -ane $H_{(2n+2)} = H_{(2(4)+2)}$ is 10, so the formula is C_4H_{10}
   2. pent- = 5 C in the chain, for -ane $H_{(2n+2)} = H_{(2(5)+2)}$ is 12, so the formula is C_5H_{12}
   3. hex- = 6 C in the chain, for -ane $H_{(2n+2)} = H_{(2(6)+2)}$ is 14, so the formula is C_6H_{14}
   4. hept- =7 C in the chain, for -ane $H_{(2n+2)} = H_{(2(7)+2)}$ is 16, so the formula is C_7H_{16}
   5. oct- = 8 C in the chain, for -ane $H_{(2n+2)} = H_{(2(8)+2)}$ is 18, so the formula is C_8H_{18}
   6. non- = 9 C in the chain, for -ane $H_{(2n+2)} = H_{(2(9)+2)}$ is 20, so the formula is C_9H_{20}
   7. dec- = 10 C in the chain, for -ane $H_{(2n+2)} = H_{(2(10)+2)}$ is 22, so the formula is $C_{10}H_{22}$

2-2   1. 4 C is but- and 10 H = 2(4) + 2, so the ending is -ane; butane _____
   2. 5 C is pent- and 12 H = 2(5) + 2, so the ending is -ane; pentane _____
   3. 6 C is hex- and 14 H = 2(6) + 2, so the ending is -ane; hexane _____

4. 7 C is hept- and 16 H = 2(7) + 2, the so ending is -ane; heptane

5. 8 C is oct- and 18 H = 2(8) + 2, so the ending is -ane; octane

6. 9 C is non- and 20 H = 2(9) + 2, so the ending is -ane; nonane

7. 10 C id dec-_and_22 H = 2(10) + 2, so the ending is -ane; decane

2-3

1. $H-C-C-C-C-H$ $CH_3CH_2CH_2CH_3$

2. $H-C-C-C-C-C-H$ $CH_3CH_2CH_2CH_2CH_2CH_3$

3. $H-C-C-C-C-C-C-H$ $CH_3CH_2CH_2CH_2CH_2CH_3$

4. $H-C-C-C-C-C-C-C-H$ $CH_3CH_2CH_2CH_2CH_2CH_2CH_3$

5. $H-C-C-C-C-C-C-C-C-H$ $CH_3CH_2CH_2CH_2CH_2CH_2CH_2CH_3$

2-4   1. $CH_3(CH_2)_2CH_3$ 2. $CH_3(CH_2)_3CH_3$ 3. $CH_3(CH_2)_4CH_3$ 4. $CH_3(CH_2)_5CH_3$ 5. $CH_3(CH_2)_7CH_3$

2-5   1. ⟍⟋⟍⟋⟍⟋   2. ⟍⟋⟍   3. ⟍⟋   4. ⟍⟋⟍
        5. ⟍⟋⟍⟋⟍⟋⟍

2-6
1. butane, $CH_3(CH_2)_2CH_3$, ⟍⟋⟍
2. heptane, $CH_3(CH_2)_5CH_3$, ⟍⟋⟍⟋⟍⟋
3. decane, $CH_3(CH_2)_8CH_3$, ⟍⟋⟍⟋⟍⟋⟍⟋⟍
4. pentane, $CH_3(CH_2)_3CH_3$, ⟍⟋⟍⟋
5. hexane, $CH_3(CH_2)_4CH_3$, ⟍⟋⟍⟋⟍

2-7
1. C_4H_{10}, butane, $CH_3(CH_2)_2CH_3$
2. C_7H_{16}, heptane, $CH_3(CH_2)_5CH_3$
3. $C_{10}H_{22}$, decane, $CH_3(CH_2)_8CH_3$
4. C_9H_{20}, nonane, $CH_3(CH_2)_7CH_3$
5. C_6H_{14}, hexane, $CH_3(CH_2)_4CH_3$

2-8
1. C_6H_{14}, 86.17 g/mol; C_8H_{18}, 114.94 g/mol; both have only dispersion forces, so octane (with the greater number of electrons) has the higher boiling point.

2. C_4H_{10}, 58.12 g/mol; C_9H_{20}, 128.3 g/mol; both have only dispersion forces, so nonane (with the greater number of electrons) has the higher boiling point.

3. CH_4, 16.04 g/mol; $C_{10}H_{22}$, 142.28 g/mol; both have only dispersion forces, so decane (with the greater number of electrons) has the higher boiling point.

4. C_8H_{18}, 114.94 g/mol; $C_{10}H_{22}$, 142.28 g/mol; both have only dispersion forces, so decane (with the greater number of electrons) has the higher boiling point.

5. C_3H_8, 44.09 g/mol; C_5H_{12}, 72.15 g/mol; both have only dispersion forces, so pentane with (the greater number of electrons) has the higher boiling point.

2-9 1. 2,3-dimethylpentane 2. 4-ethyl-2,3-dimethylheptane 3. 2,2,3,3-tetramethylbutane
4. 3,3,4,4-tetramethylhexane 5. 2-methylnonane

2-10 1.

2.

3.

4.

5.

2-11 1. $C_{11}H_{24}$ 2. $C_{14}H_{30}$ 3. $C_{12}H_{26}$ 4. C_9H_{20} 5. $C_{12}H_{26}$

2-12 1. $CH_3CH(CH_3)CH(CH_3)(CH_2)_5CH_3$

2. $CH_3C(CH_3)_2CH_2CH(CH_2CH_2CH_3)(CH_2)_4CH_3$

3. $CH_3CH_2CH_2CH(CH_2CH_3)\ C(CH_3)_2(CH_2)_3CH_3$

4. $CH_3C(CH_3)_2\ CH_2CH(CH_3)CH_2CH_3$

5. $CH_3CH(CH_3)CH(CH_3)CH(CH_2CH_3)(CH_2)_3CH_3$

2-13 1. 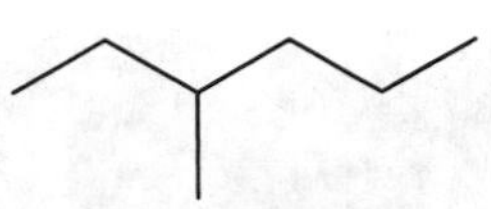 2. 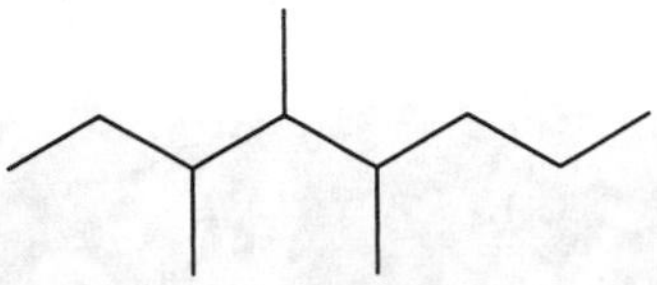 3.

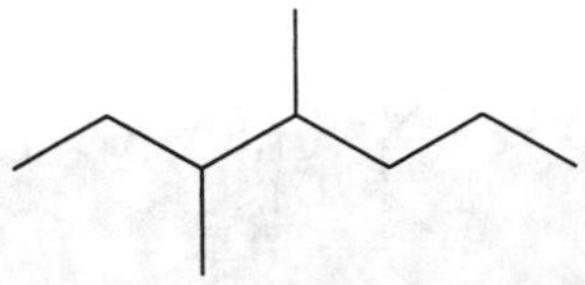

4. 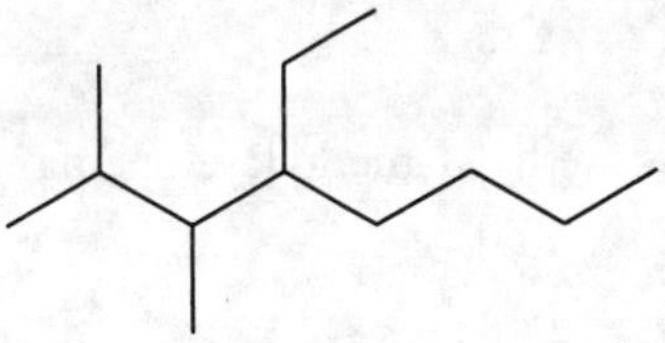 5.

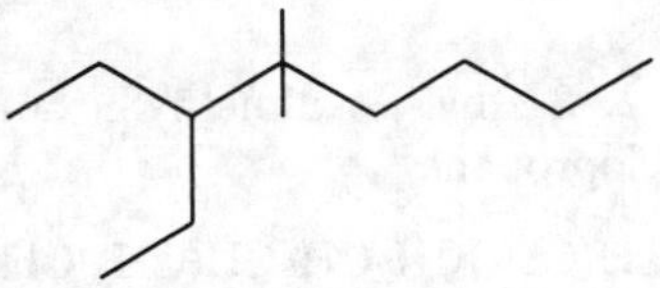

2-14

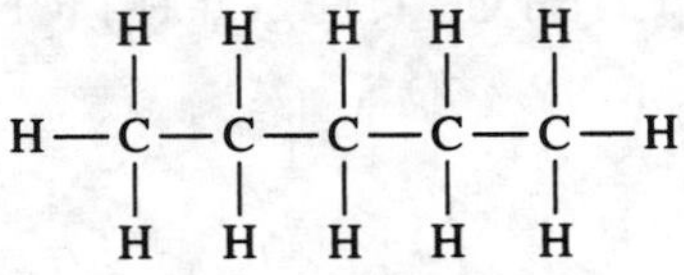

A-pentane

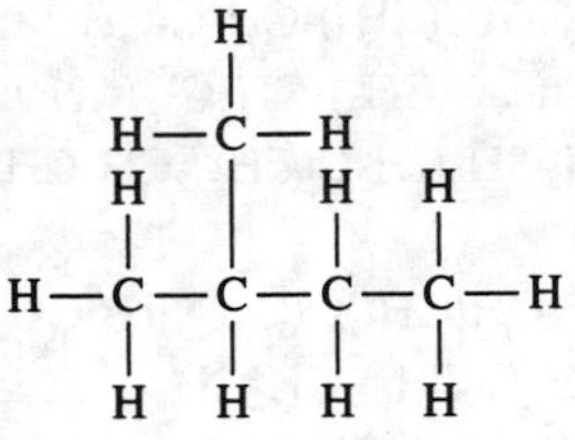

2-methylbutane

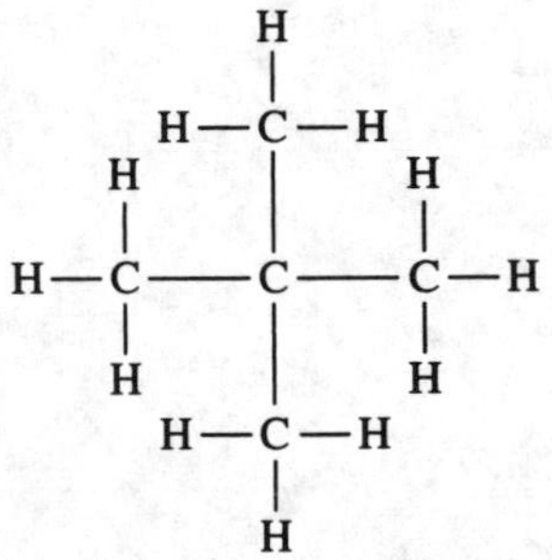

2, 3-dimethylpropane

2-15 1. 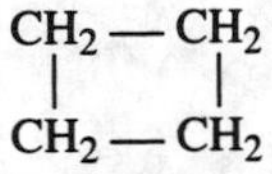 C_4H_8

2. 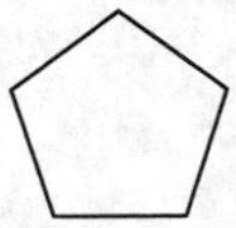 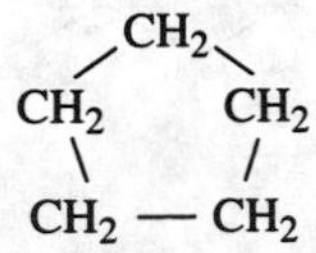 C_5H_{10}

3. 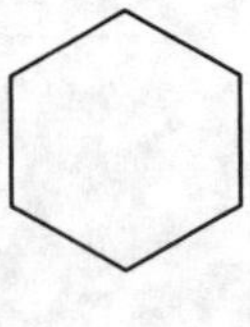 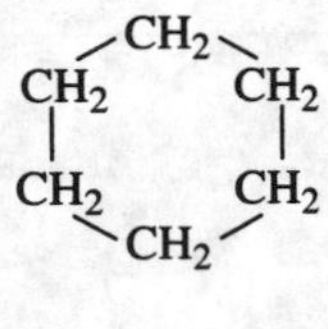 C_6H_{12}

4. 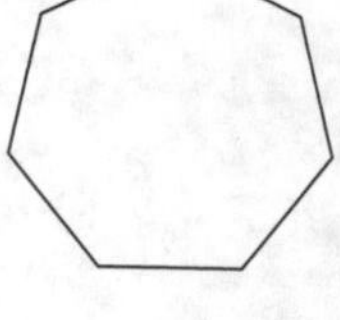 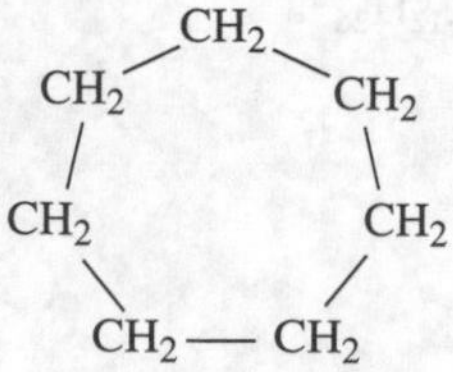 C_7H_{14}

5. 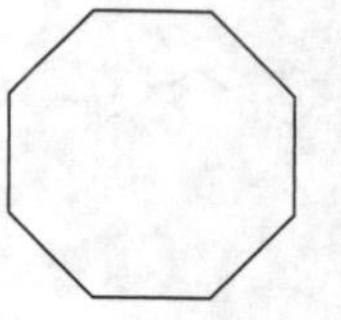 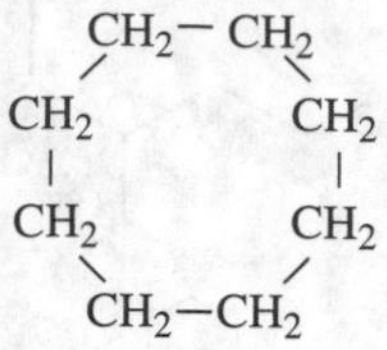 C_8H_{16}

2-16 1. 1,3-dimethylcyclohexane   2. 2-ethyl-1,4-dimethylhexane   3. 2-ethyl-1-methylcyclopentane
4. cyclobutane   5. 1-ethylcyclopropane

2-17 1. 1,4-dimethylcyclohexane, $CH(CH_3)CH_2CH_2CH(CH_3)CH_2CH_2$
2. 2,2-dimethylbutane, $CH_3C(CH_3)_2CH_2CH_3$
3. 2,2,4,4-tetramethylpentane, $CH_3C(CH_3)_2CH_2C(CH_3)_2CH_3$
4. 1,1,3,5-tetramethylcyclohexane, $C(CH_3)_2CH_2CH(CH_3)CH_2CH(CH_3)CH_2$
5. 5,5-dimethyldecane, $CH_3CH_2CH_2CH_2C(CH_3)_2CH_2CH_2CH_2$ CH_2CH_3 or $CH_3(CH_2)_3C(CH_3)_2(CH_2)_4CH_3$
6. pentane, $CH_3(CH_2)_3CH_3$
7. 2-ethyl-1-methylcyclohexane, $CH(CH_3)C(CH_2CH_3)CH_2CH_2CH_2$
8. 3,3-dimethylpentane, $CH_3CH_2C(CH_3)_2CH_2CH_3$
9. 3-methylpentane, $CH_3CH_2CH(CH_3)CH_2CH_3$
10. 2,4,5-trimethylheptane, $CH_3CH(CH_3)CH_2CH(CH_3)CH(CH_3)CH_2CH_3$

2-18 1.

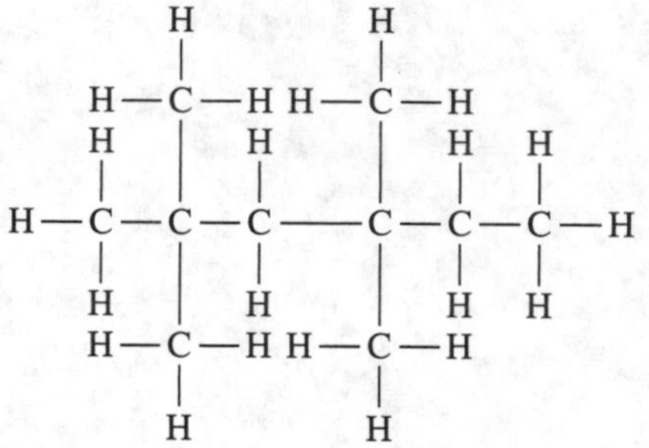

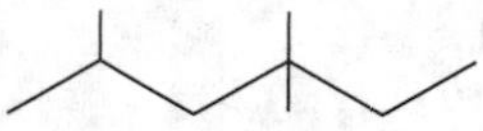

2.

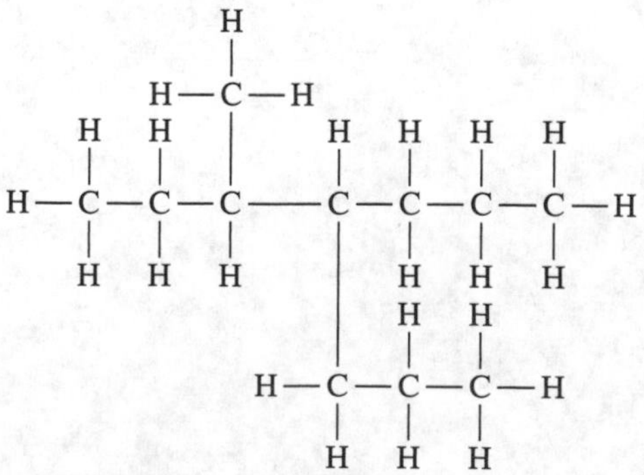

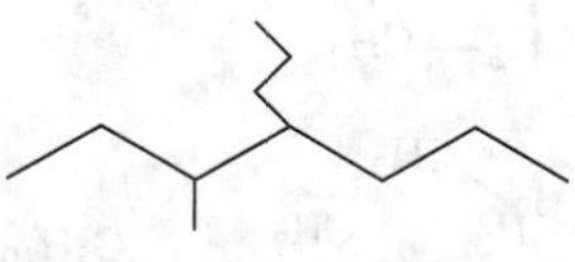

3.

```
      H   CH₃              
      |   |                
  H   C                H   
   \ / \              |    
H—C     C—CH₃              
   |     |                 
H—C     C—H               
   |     |                 
   H     H                 
      \ /                  
       C                   
      / \                  
     H   H                 
```

(cyclohexane ring with two methyl substituents, skeletal form)

4.

```
       H                H
       |                |
   H—C—H            H—C—H
       |                |
   H   H            H   H   H   H
   |   |   |        |   |   |   |
H—C—C—C————C—C—C—C—C—H
   |   |   |        |   |   |   |
   H   H   |        H   H   H   H
       H—C—C—H
           |   |
           H   H
```

5.

```
       H
       |
   H—C—H
       |
   H   H        H   H   H   H
   |   |        |   |   |   |
H—C—C————C—C—C—C—C—H
   |   |        |   |   |   |
   H   |        H   H   H   H
   H—C—C—H
       |   |
       H   H
```

6.

```
       H
       |
   H—C—H
       |
   H   H   H   H   H
   |   |   |   |   |
H—C—C—C—C—C—C—H
   |   |   |   |   |
   H   H   H   H   H
```

7.

```
       H            H
       |            |
   H—C—H    H—C—H
       |            |
   H   H   |        H   H
   |   |   |        |   |   |   |
H—C—C—C—C————C—C—C—C—H
   |   |   |        |   |   |   |
   H   H   H        H   H   H
       H—C—H
           |
           H
```

8.

```
       H   H
       |   |
   H—C—C—H
       |   |
   H   H   |   H
   |   |   |   |   |
H—C—C—C————C—C—H
   |   |   |   |   |
   H   H   |   H   H
       H—C—C—H
           |   |
           H   H
```

9.

10.

3 Alkenes

3-1   1. 1-butene   2. 2-methyl-1-butene   3. 3-methyl-2-pentene   4. 3,4-dimethyl-3-hexene
5. 3-ethyl-4,6-dimethyl-1-heptene   6. 2,2,4-trimethyl-3-hexene   7. 1-cyclopentene
8. 4-ethyl-2,4-dimethyl-2-hexene   9. 3-methyl-1-butene   10. 1,4-pentadiene

3-2   1.

2.

3.

4.

5.

6.

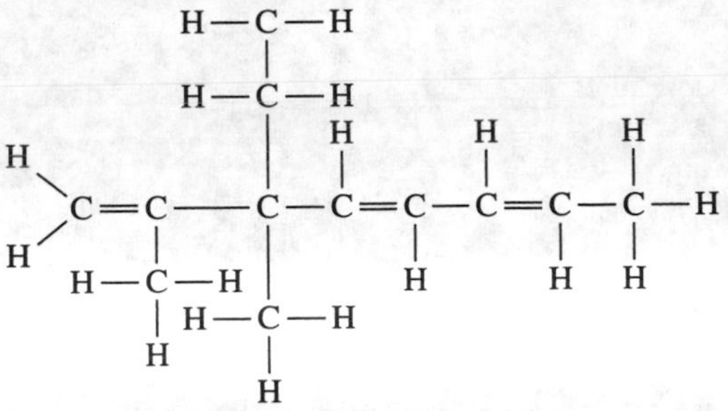

7.

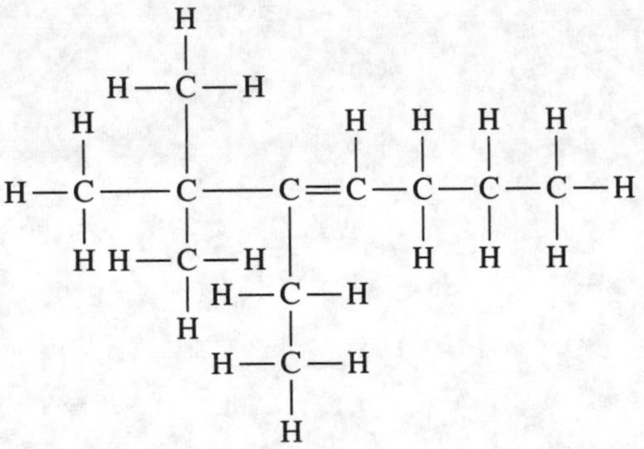

8.

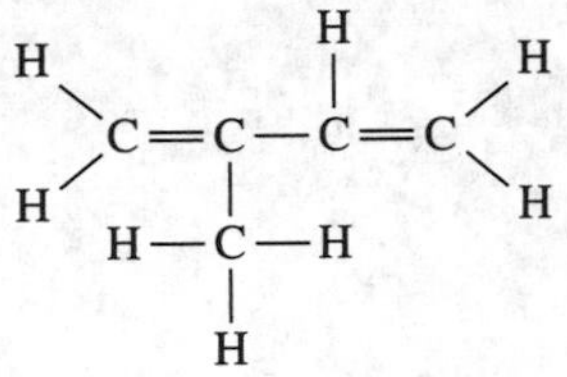

9.

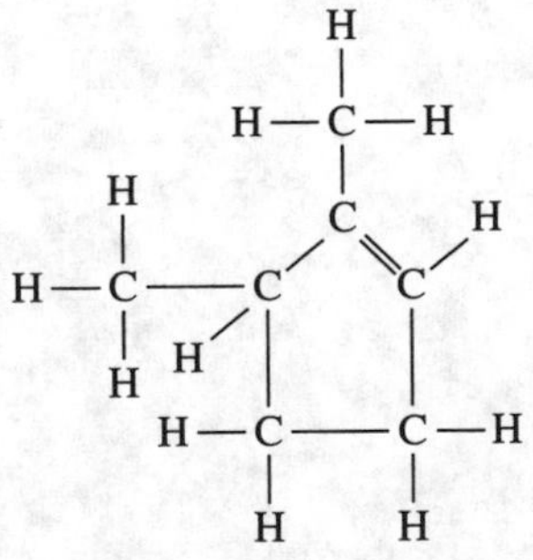

3-3

1. $CH_3CH_2CH=CHCH_2CH_3$

2. $CH_3(CH_3)C=C(CH_3)CH_2CH_3$

3. $CH_3HC=C(CH_3)CH_2CH_3$

4. $HC=CHCH_2HC=CHCH_2$

5. $(CH_3)C=CHCH(CH_3)CH_2CH_2$

6. $H_2C=C(CH_3)C(CH_3)_2HC=CHHC=CHCH_3$

7. $CH_3C(CH_3)_2(CH_2CH_3)C=CHCH_2CH_2CH_3$

8. $H_2C=C(CH_3)HC=CH$

9. $(CH_3)C=CHCH_2CH_2CH(CH_3)$

3-4

1.

2. 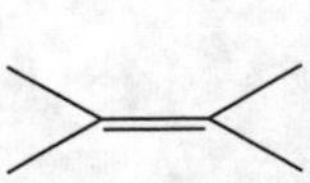   or

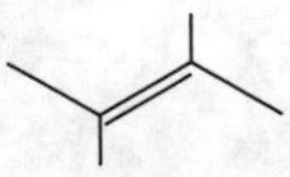

3.

4.

5.

6.

7.

8.

9.

3-5   1. Br   CH₃
               Cl
               Cl

2. H   H
   H   F
   H
   H   NO₂
   H   H

or

   H   H
   H—C   C—F
   H—C   C—NO₂
   H   H

3-6   1. 5,5-dibromo-2-heptadiene
          2. 1,4-dimethyl-1,4-cyclohexadiene

3-7   1. Br
               Br

          2. Br
                  Br

          3. Br
                  NO₂

4.

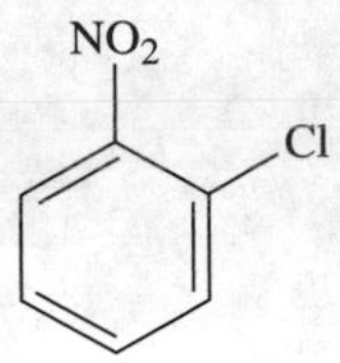

5. 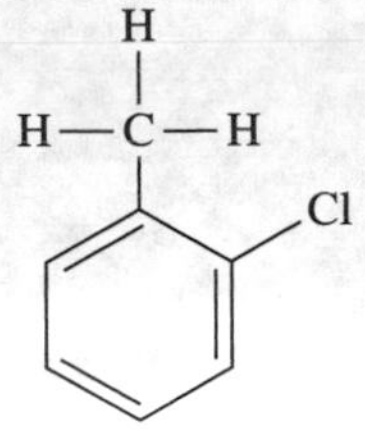 H—C—H

3-8

1. 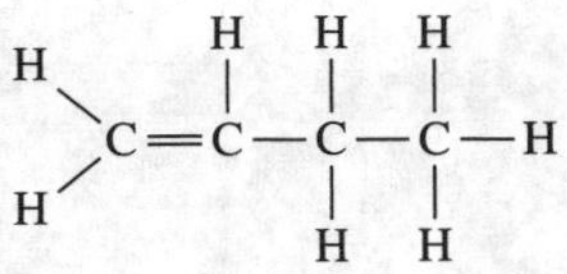 eleven σ
one π

2. 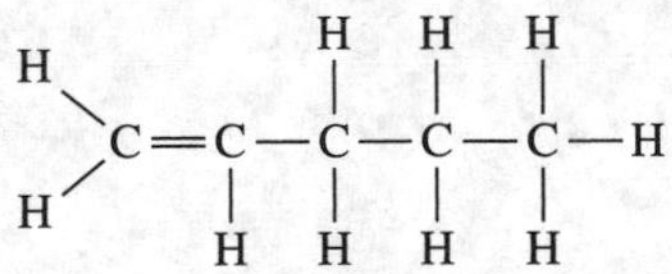 fourteen σ
one π

3. 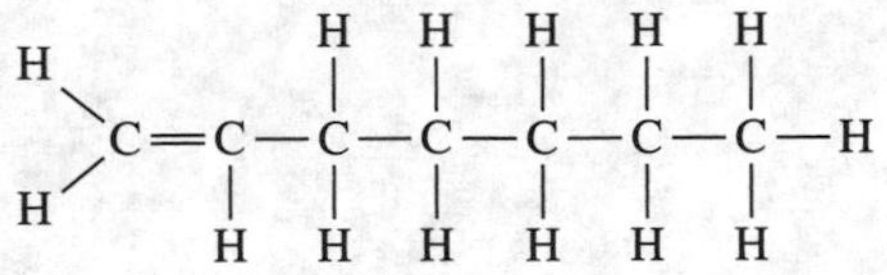 twenty σ
one π

4. 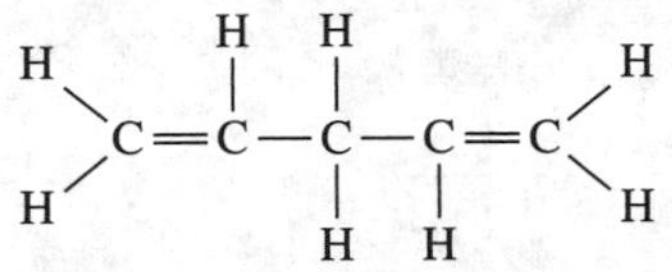 twelve σ
two π

5. 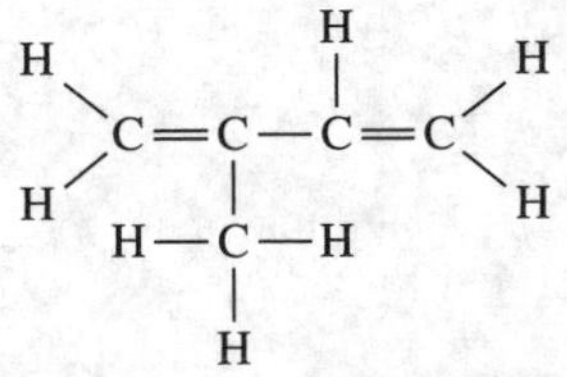 twelve σ
two π

3-9

1. 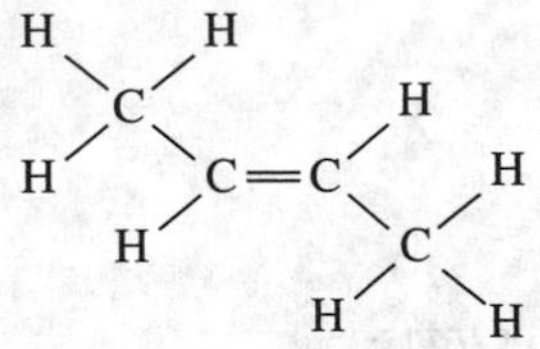 *trans*-2-butene

2. 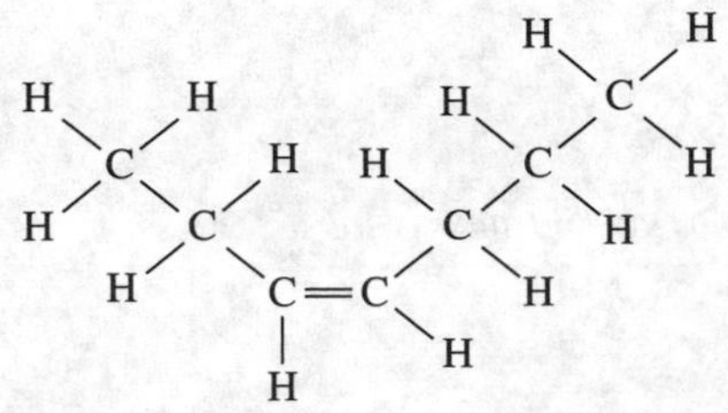 *cis*-3-heptene

3.

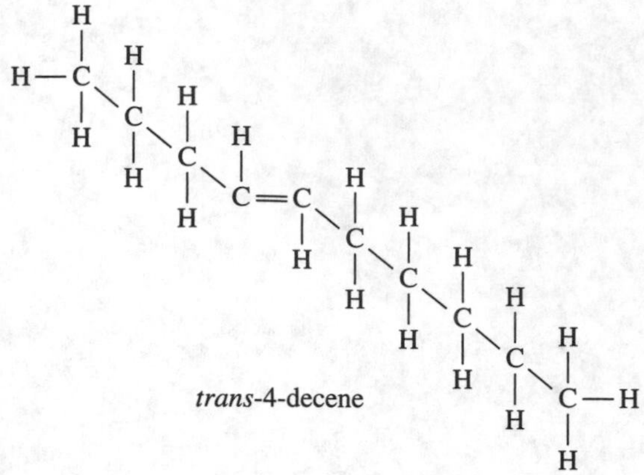

trans-4-decene

4.

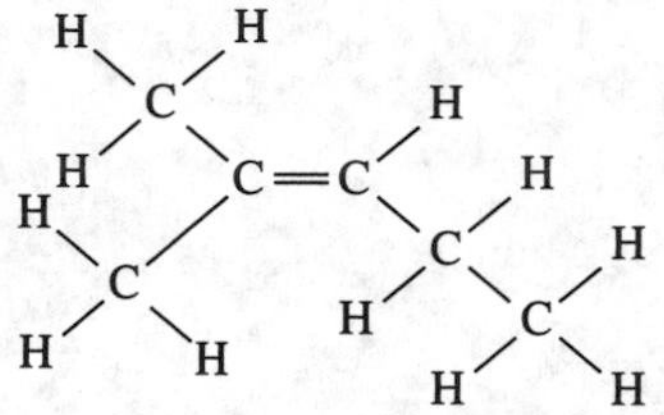

2-methyl-*trans*-2-pentene

5.

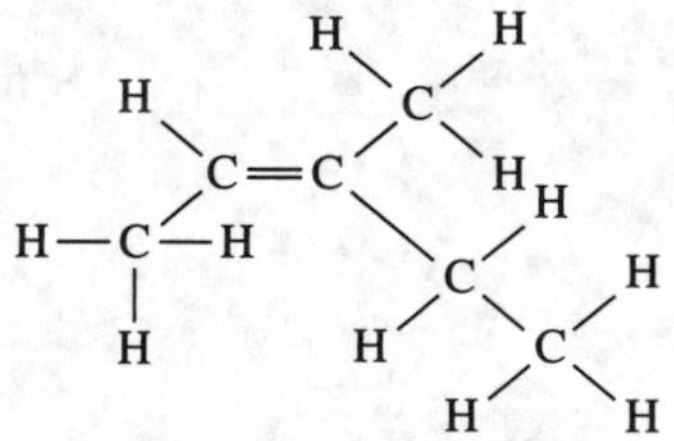

3-methyl-*cis*-2-pentene

3-10   1.

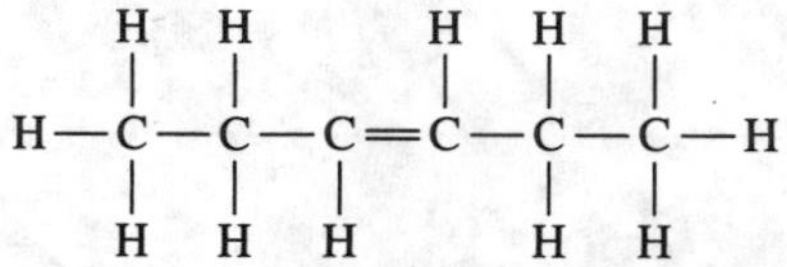

can also be *cis*- or *trans*-

2.

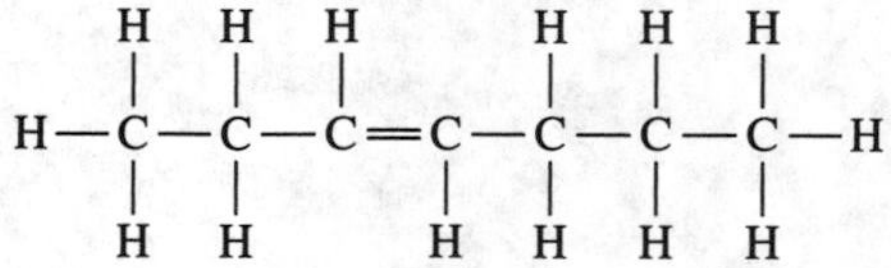

can also be *cis*- or *trans*-

3.

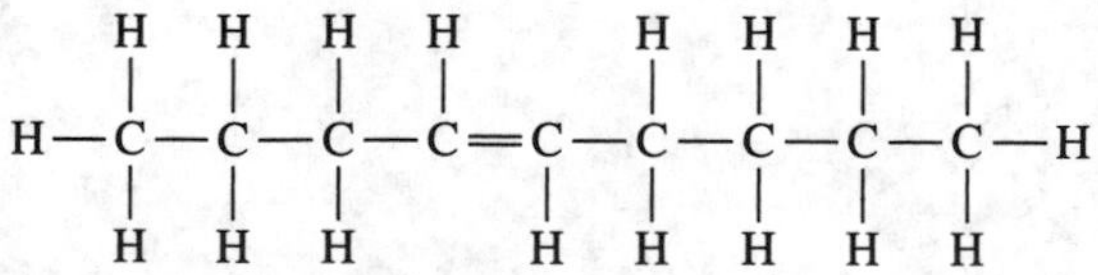

can also be *cis*- or *trans*-

4.

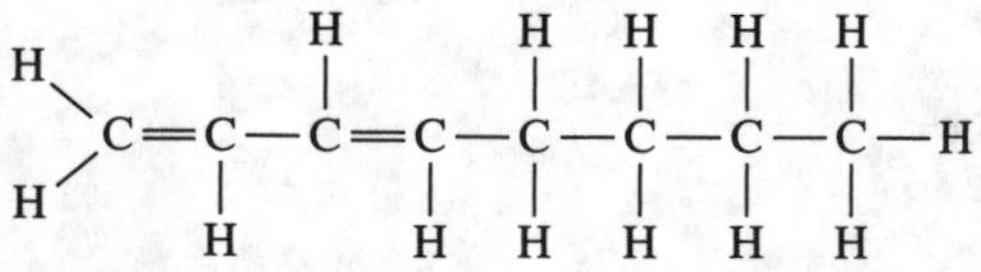

can also be *cis*- or *trans*-

5.

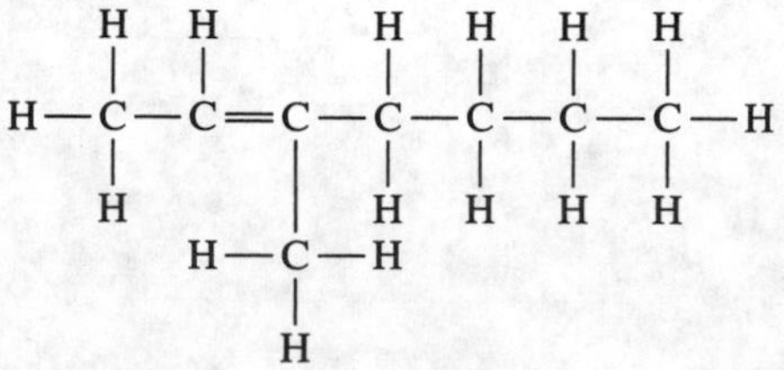

can also be *cis*- or *trans*-

6.

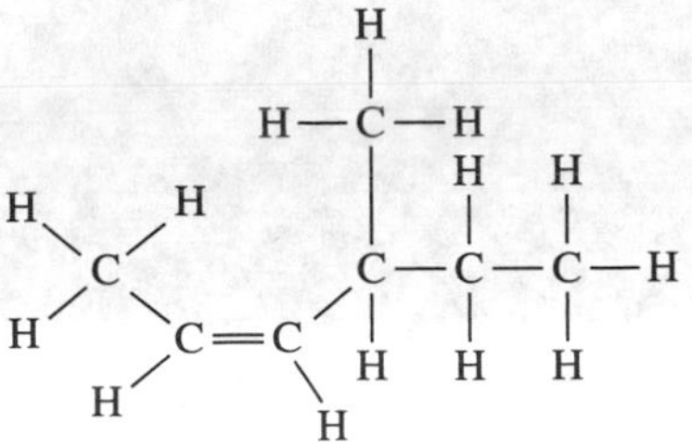

3-11   1. $CH_3CH_2HC{=}CHCH_2CH_3$
2. $CH_3CH_2HC{=}CHCH_2CH_2CH_3$
3. $CH_3CH_2CH_2HC{=}CH(CH_2)_3CH_3$
4. $H_2C{=}CHHC{=}CH(CH_2)_3CH_3$
5. $CH_3HC{=}C(CH_3)(CH_2)_3CH_3$
6. $CH_3HC{=}CHCH(CH_3)CH_2CH_3$

3-12   1. C_5H_{10} 2. C_8H_{10} 3. $C_7H_{12}Cl$ 4. $C_8H_{16}Br_2$ 5. C_4H_8

3-13   1. 1,3-dichloronapthalene   2. 3,3-dimethyl-1-butene   3. 2-methyl-2-butene
4. 2-methyl-1,4-pentadiene   5. 2-pentene

3-14   1.

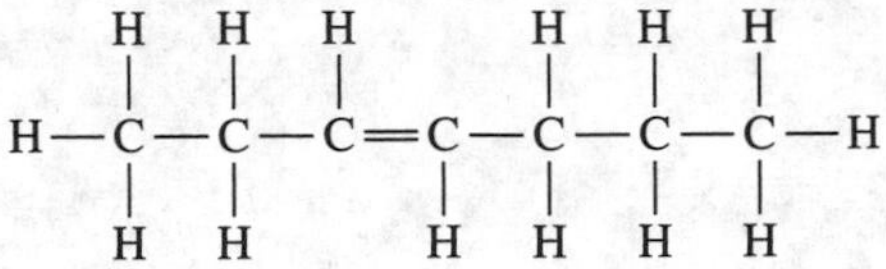

2. 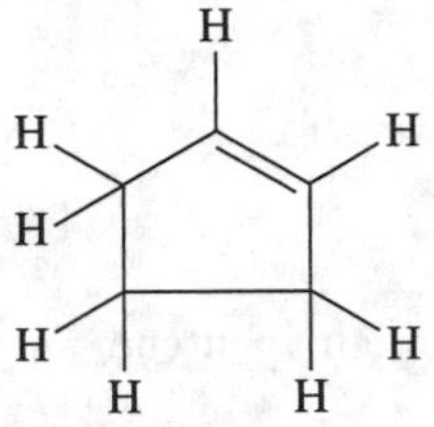   or

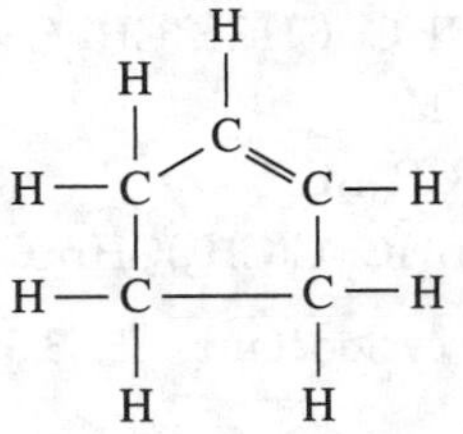

3.

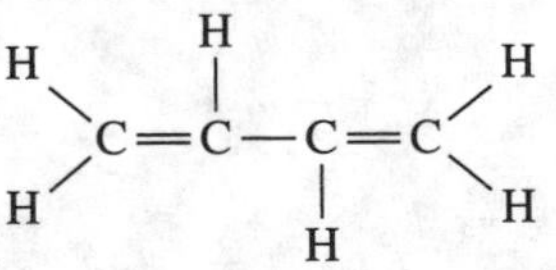

4. 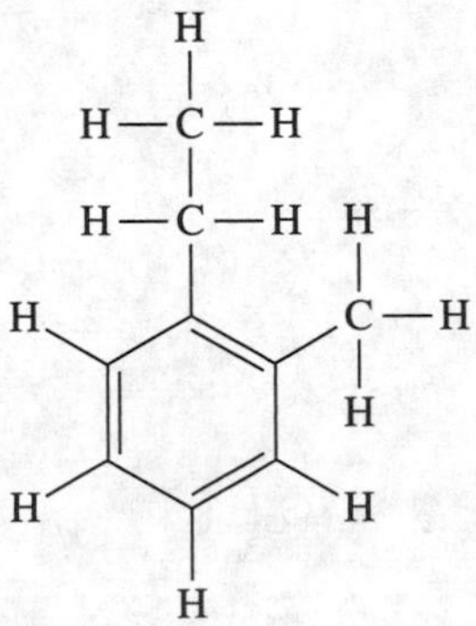   or

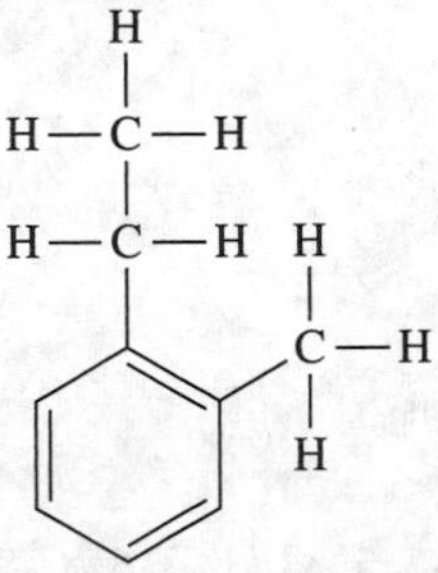

5.

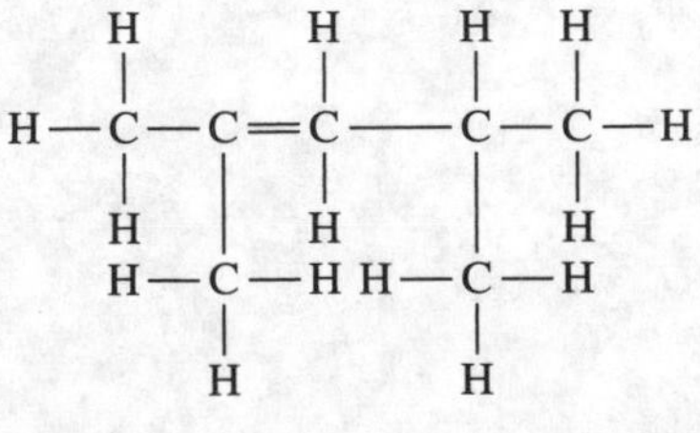

3-15   1.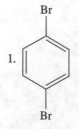

2.

3.

4.

5.

3-16   a. C_5H_{10}, 2-pentene, $CH_3HC=CHCH_2CH_3$

b. C_9H_{16}, 6-methyl-2,4-octdiene, $CH_3HC=CHHC=CHCH(CH_3)CH_2CH_3$

c. C_9H_{16}, 2,3-nondiene, $CH_3HC=C=CHCH_2CH_3$

d. C_6H_{12}, 3,3-dimethyl-1-butene, $H_2C=CHC(CH_3)_2CH_3$

e. $C_6H_{11}Br$, 4-bromo-3-methyl-2-pentene, $CH_3HC=C(CH_3)CHBrCH_3$

3-17   1. hexane   2. 3-heptene   3. 4-nonene   4. cyclooctane   5. 3,3,4,4-tetramethylpentene

4   Alkynes

4-1   1. 1-octyne, C_8H_{14}

2. 5-ethyl-2-methyl-3-heptyne, $C_{10}H_{18}$

3. 2-butyne, C_4H_6

4. 1-propyne, C_3H_4

5. 3-hexyne, C_6H_{10}

6. 3-ethyl-4-methyl-1-pentyne, C_8H_{14}

7. cyclohexyne, C_6H_8

8. 4,4-dimethyl-2-hexyne, C_8H_{14}

9. 4-ethyl-5,5-dimethyl-2-hexyne, $C_{10}H_{18}$

10. 3,6-dimethyl-4-nonyne, $C_{11}H_{20}$

4-2   1. C_2H_2 2. C_3H_4 3. C_4H_6 4. C_5H_8 5. C_6H_{10} 6. $C_{10}H_{10}$ 7. C_9H_{16} 8. C_6H_{10}
9. $C_{10}H_{18}$ 10. C_8H_{14}

4-3   1. $H—C\equiv C—H$

2.

3.

```
    H           H
    |           |
H — C — C ≡ C — C — H
    |           |
    H           H
```

4.

```
        H   H   H
        |   |   |
H — C ≡ C — C — C — C — H
        |   |   |
        H   H   H
```

5.

```
        H   H   H   H
        |   |   |   |
H — C ≡ C — C — C — C — C — H
        |   |   |   |
        H   H   H   H
```

6.

```
        H   H
        |   |
H — C ≡ C — C — C — H
        |   |
        |   H
      H |   | H
       \ C = C /
        ‖     ‖
       / C — C \
      H |     | H
        C = C
        |
        H
```

7.

```
    H   H           H   H   H   H
    |   |           |   |   |   |
H — C — C — C ≡ C — C — C — C — C — H
    |   |           |   |   |   |
    H   H           H   |   H   H
                        |
                      H — C — H
                        |
                        H
```

8.

```
    H           H   H   H
    |           |   |   |
H — C — C ≡ C — C — C — C — H
    |           |   |   |
    H           H   H   H
```

9.

```
            H
            |
          H — C — H
            |
    H   H   |               H   H   H
    |   |   |               |   |   |
H — C — C — C — C ≡ C — C — C — C — H
    |   |   |               |   |   |
    H   H   |               H   H   H
          H — C — H
            |
            H
```

10.

```
    H   H   H               H   H   H
    |   |   |               |   |   |
H — C — C — C — C ≡ C — C — C — C — H
    |   |   |               |   |   |
    H   H   H               H   H   H
```

4-4   1.   H — C ≡ C — H

2.

```
                    H
                   ⁝
H — C ≡ C — C ⋯⋯ H
                  ╲
                   H
```

3.

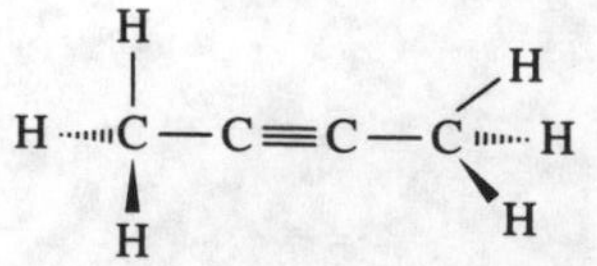

4.

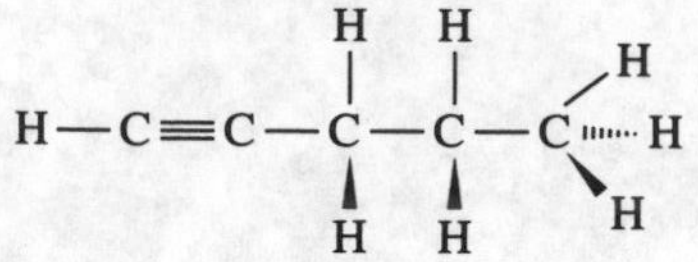

5.

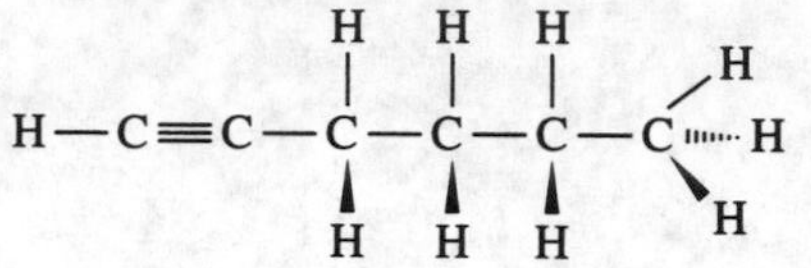

6.

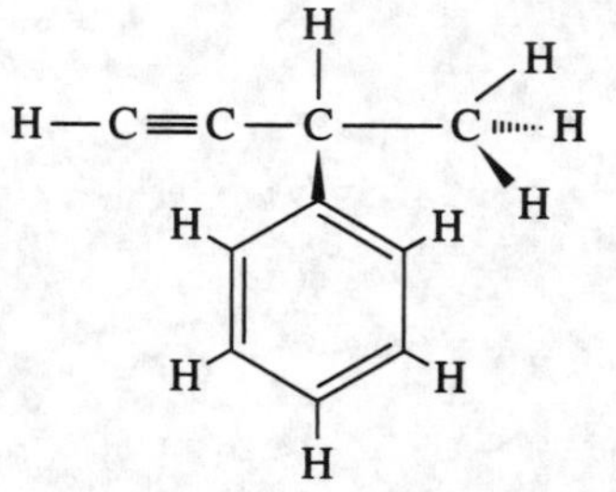

7.

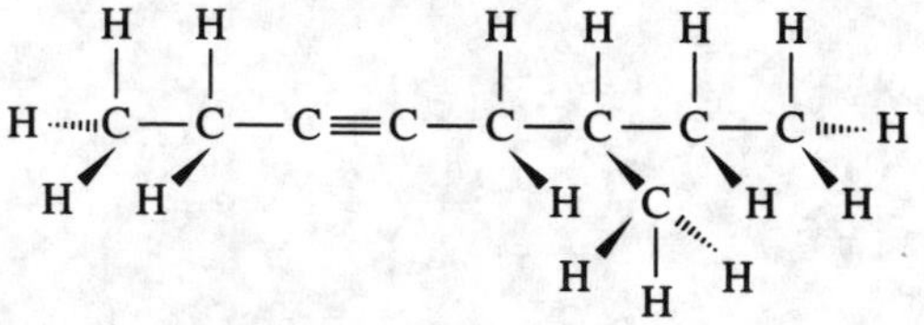

8.

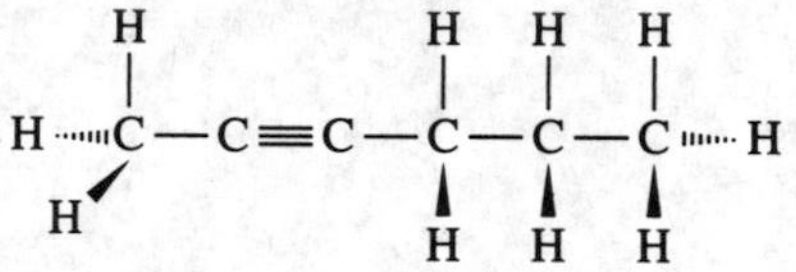

9.

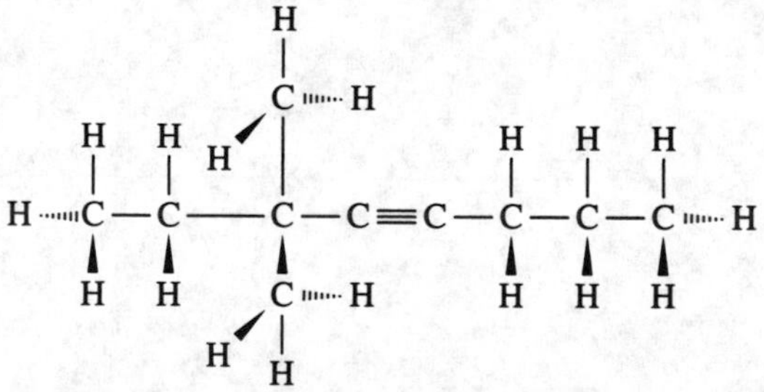

10.

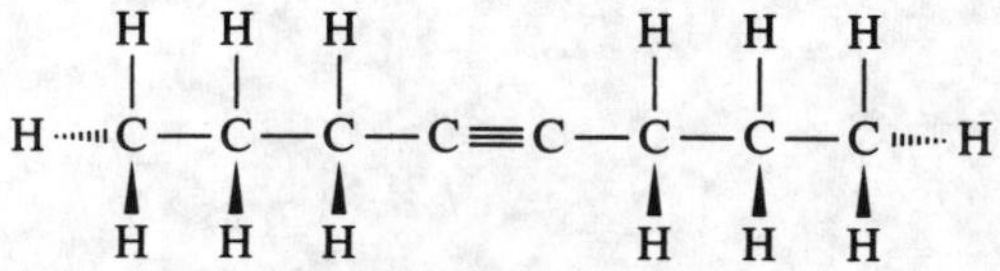

4-5

1. $HC\equiv CH$

2. $HC\equiv CCH_3$

3. $CH_3C\equiv CCH_3$

4. $HC\equiv CCH_2CH_2CH_3$

5. $HC\equiv CCH_2CH_2CH_2CH_3$

6. $HC{\equiv}C(C_6H_5)CH_3$

7. $CH_3CH_2C{\equiv}CCH_2CH(CH_3)CH_2CH_3$

8. $CH_3C{\equiv}CCH_2CH_2CH_3$

9. $CH_3CH_2C(CH_3)_2C{\equiv}CCH_2CH_2CH_3$

10. $CH_3CH_2CH_2C{\equiv}CCH_2CH_2CH_3$

4-6

1.

2.

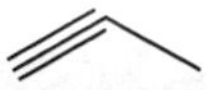

3.

4.

5.

6.

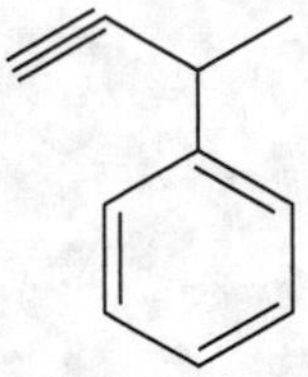

7.

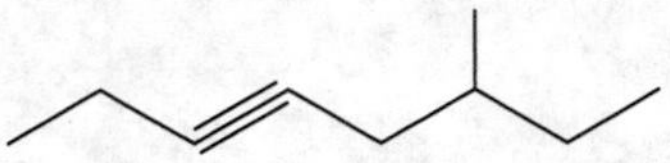

8.

9.

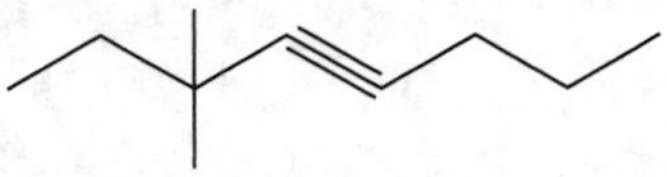

10.

4-7

(In each case, there are many other possibile isomers. Only two are shown.)

1. Isomer 1

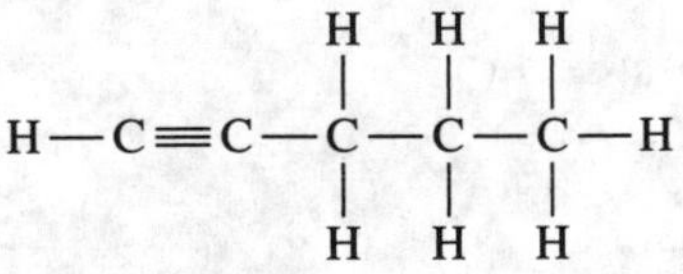

1-pentyne

$HC{\equiv}CCH_2CH_2CH_3$

Isomer 2

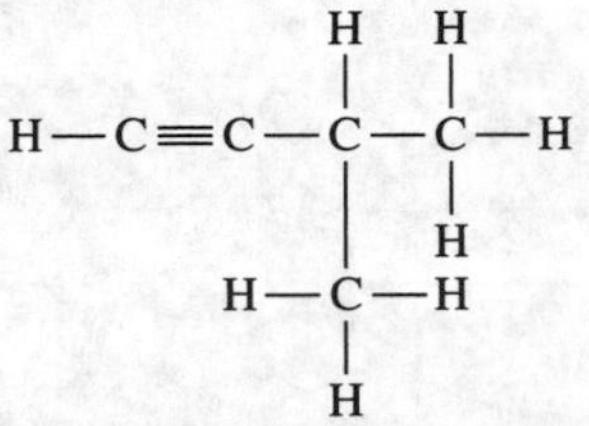

3-methyl-1-butyne

$HC{\equiv}CCH(CH_3)CH_2CH_3$

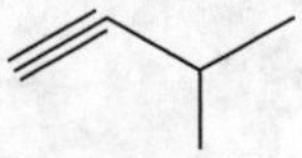

2. Isomer 1

$$H-C\equiv C-\overset{\overset{\displaystyle H}{|}}{\underset{\underset{\displaystyle H}{|}}{C}}-\overset{\overset{\displaystyle H}{|}}{\underset{\underset{\displaystyle H}{|}}{C}}-\overset{\overset{\displaystyle H}{|}}{\underset{\underset{\displaystyle H}{|}}{C}}-\overset{\overset{\displaystyle H}{|}}{\underset{\underset{\displaystyle H}{|}}{C}}-\overset{\overset{\displaystyle H}{|}}{\underset{\underset{\displaystyle H}{|}}{C}}-H$$

1-heptyne
$HC\equiv CCH_2CH_2CH_2CH_2CH_3$

Isomer 2

$$H-C\equiv C-\overset{\overset{\displaystyle H}{|}}{\underset{|}{C}}-\overset{\overset{\displaystyle H}{|}}{\underset{|}{C}}-\overset{\overset{\displaystyle H}{|}}{\underset{\underset{\displaystyle H}{|}}{C}}-H$$

3,4-dimethyl-1-pentyne
$HC\equiv CHC(CH_3)HC(CH_3)CH_3$

3. Isomer 1

$$H-\overset{\overset{\displaystyle H}{|}}{\underset{\underset{\displaystyle H}{|}}{C}}-\overset{\overset{\displaystyle H}{|}}{\underset{\underset{\displaystyle H}{|}}{C}}-C\equiv C-\overset{\overset{\displaystyle H}{|}}{\underset{\underset{\displaystyle H}{|}}{C}}-\overset{\overset{\displaystyle H}{|}}{\underset{\underset{\displaystyle H}{|}}{C}}-\overset{\overset{\displaystyle H}{|}}{\underset{\underset{\displaystyle H}{|}}{C}}-\overset{\overset{\displaystyle H}{|}}{\underset{\underset{\displaystyle H}{|}}{C}}-\overset{\overset{\displaystyle H}{|}}{\underset{\underset{\displaystyle H}{|}}{C}}-\overset{\overset{\displaystyle H}{|}}{\underset{\underset{\displaystyle H}{|}}{C}}-H$$

3-decyne
$CH_3CH_2C\equiv C(CH_2)_5CH_3$

Isomer 2

2,5,5-trimethyl-3-heptyne
$CH_3HC(CH_3)C\equiv CC(CH_3)_2CH_2CH_3$

4. Isomer 1

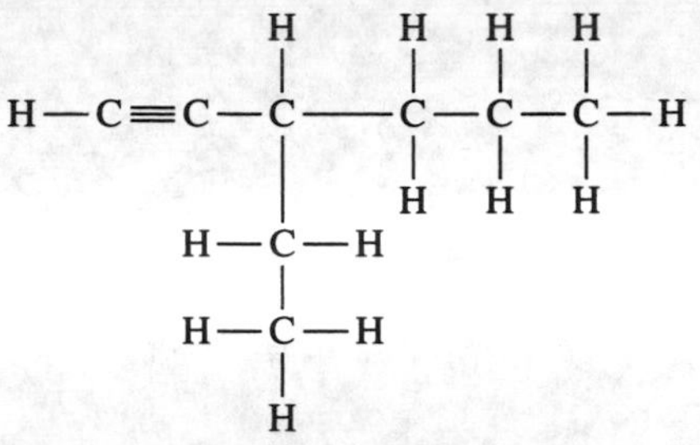

3-ethyl-1-hexyne

$HC{\equiv}CHC(CH_2CH_3)CH_2CH_2CH_3$

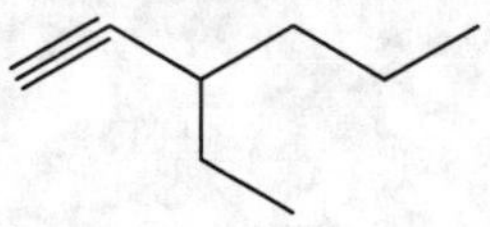

Isomer 2

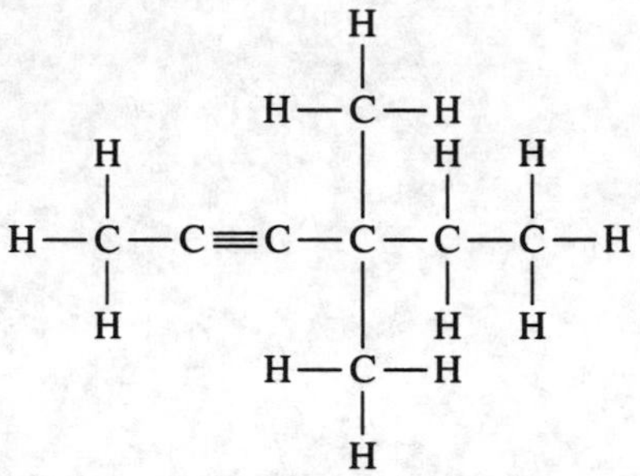

4,4-dimethyl-2-hexyne

$CH_3C{\equiv}CC(CH_3)_2CH_2CH_3$

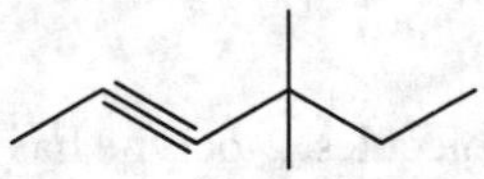

5. Isomer 1

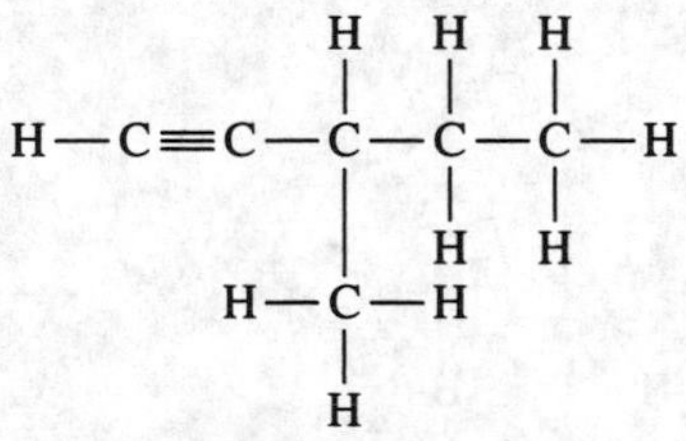

3-methyl-1-pentyne

$HC{\equiv}CHC(CH_3)CH_2CH_3$

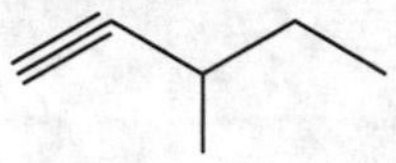

Isomer 2

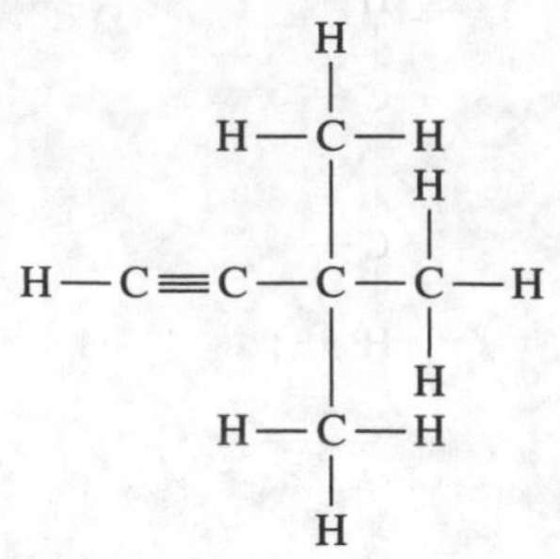

3,3-dimethyl-1-butyne
$HC\equiv CC(CH_3)_2CH_3$

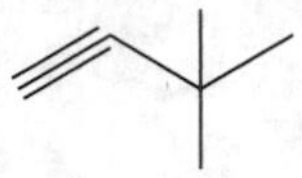

4-8   1. $C_{10}H_{17}$
4,5,5-trimethyl-2-heptyne
$CH3C\equiv CCH(CH_3)C(CH_3)_2CH_2CH_3$

2. C_9H_{12}
7-methyl-1,4-octadiyne
$HC\equiv CCH_2C\equiv CCH_2HC(CH_3)CH_3$

3. $C_{13}H_{16}$
7-phenyl-2-heptyne
$CH_3C\equiv CCH_2CH_2CH_2CH_2(C_6H_5)$

4. $C_{19}H_{27}$
7-ethyl-6,8-dimethyl-3-phenyl-4-nonyne
$CH_3CH_2CH(C_6H_5)C\equiv CHC(CH_3)CH(CH_2CH_3)CH(CH_3)CH_3$

5. C_9H_{16}
3-ethyl-4-methyl-1-hexyne
$HC\equiv CCH(CH_2CH_3)CH(CH_3)CH_2CH_3$

4-9   Although both substances have only dispersion forces present between molecules, 2-octyne has a higher boiling point because of the greater number of electrons and the resulting increase in dispersion forces.

4-10   1. monosubstitued or terminal
2. disubstituted
3. monosubstitued or terminal
4. disubstituted
5. monosubstitued or terminal and disubstituted

4-11   1.

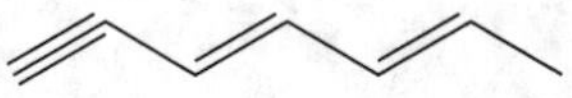

or

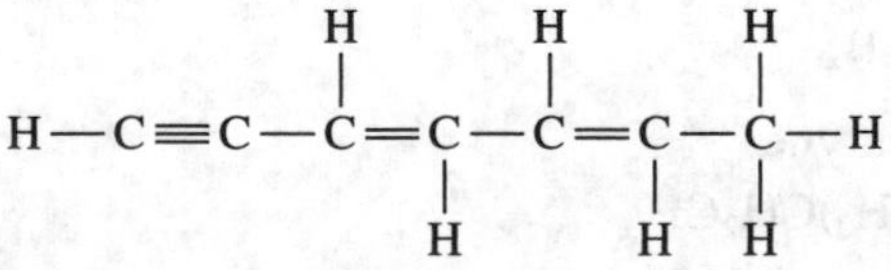

2.

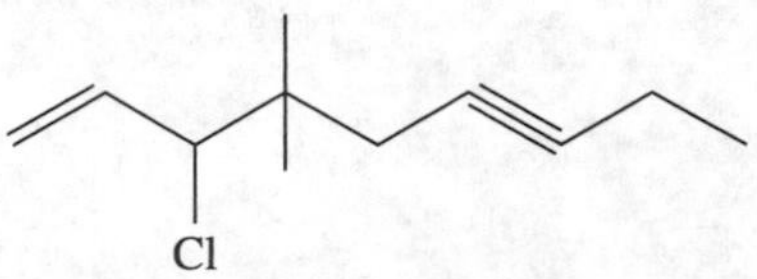

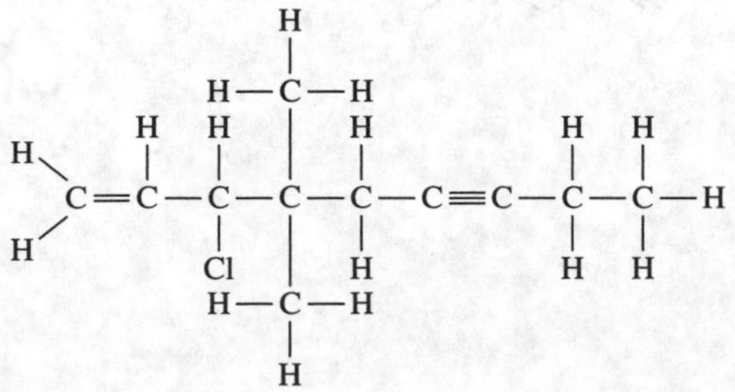

3.

4-12   1. 2,4-octdien-6-yne
2. 3,6-dimethyl-2-hepten-4-yne
3. 3,3-dimethyl-1,5-hexdiyne

5   Alcohols

5-1   1.

2.

3.

4.

5.

5-2   1. C_3H_7OH, propyl alcohol, propanol
2. C_4H_9OH, butyl alcohol, butanol
3. $C_5H_{11}OH$, pentyl alcohol, pentanol
4. $C_6H_{13}OH$, hexyl alcohol, hexanol
5. $C_8H_{17}OH$, octyl alcohol, octanol

5-3   Draw the Lewis structure and molecular geometry for each of the following alcohols.

1.
$$\begin{array}{ccc} & H & H \\ & | & | \\ H - & C - C - \ddot{O} - H \\ & | & | \\ & H & H \end{array}$$

$$\overset{H}{\underset{H}{\diagdown}} C - C \overset{\ddot{O}-H}{\underset{H}{\diagup}}$$

2.
$$H - \underset{H}{\overset{H}{C}} - \underset{H}{\overset{H}{C}} - \underset{H}{\overset{H}{C}} - \underset{H}{\overset{H}{C}} - \underset{H}{\overset{H}{C}} - \underset{H}{\overset{H}{C}} - \underset{H}{\overset{H}{C}} - \underset{H}{\overset{H}{C}} - \ddot{O} - H$$

3.
$$H - C - C - C - C - C - C - C - C \overset{\ddot{O}-H}{\diagup} \quad (\text{condensed chain structure with } H \text{ substituents})$$

3.
$$H - \underset{H}{\overset{H}{C}} - \underset{H}{\overset{H}{C}} - \underset{H}{\overset{H}{C}} - \underset{H}{\overset{H}{C}} - \ddot{O} - H$$

4. (geometry structure)
$$H - C - C - C - C \overset{\ddot{O}-H}{\diagup} \quad (\text{with } H \text{ substituents})$$

4.
$$H - \underset{H}{\overset{H}{C}} - \underset{H}{\overset{H}{C}} - \underset{H}{\overset{H}{C}} - \underset{H}{\overset{H}{C}} - \underset{H}{\overset{H}{C}} - \ddot{O} - H$$

5. (geometry structure)
$$H - C - C - C - C - C \overset{\ddot{O}-H}{\diagup} \quad (\text{with } H \text{ substituents})$$

5.
$$H - \underset{H}{\overset{H}{C}} - \underset{H}{\overset{H}{C}} - \underset{H}{\overset{H}{C}} - \underset{H}{\overset{H}{C}} - \underset{H}{\overset{H}{C}} - \underset{H}{\overset{H}{C}} - \ddot{O} - H$$

(geometry structure)
$$H - C - C - C - C - C \overset{\ddot{O}-H}{\diagup} \quad (\text{with } H \text{ substituents})$$

5-4   For the following parent alkanes, draw the condensed formula and line drawing of its alcohol with the functional group on a terminal carbon.

1. $CH_3CH_2CH_2OH$ — line drawing with OH

2. $CH_3CH_2CH_2CH_2OH$ — line drawing with OH

3. $CH_3CH_2CH_2CH_2CH_2OH$ — line drawing with OH

4. CH₃CH₂CH₂CH₂CH₂CH₂OH $\quad$

5. CH₃(CH₂)6CH₂OH

5-5
1. 1,2-butandiol
2. 3,3-pentandiol
3. 3-methyl-2,4-hexandiol
4. 4-methyl-3-hexanol
5. 3-methyl-1-butanol

5-6
1. cyclopropanol
2. 1,2-cyclobutandiol
3. phenol
4. 1,2-cyclohexandiol
5. 1,4-cyclohexandiol

5-7
1. C_3H_5OH

2. $C_4H_6(OH)_2$,

3. C_6H_5OH

4. $C_6H_{10}(OH)_2$,

5. $C_6H_{10}(OH)_2$,

5-8 For each of the following compounds, draw the structure of the compound.

1.

2.

3.

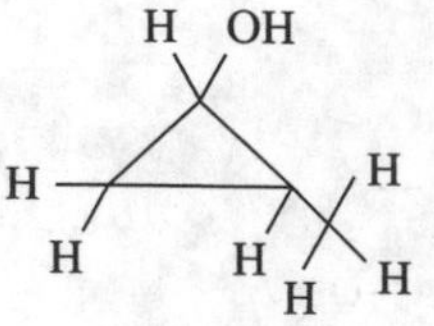

4.

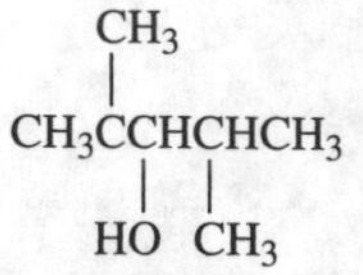

5.

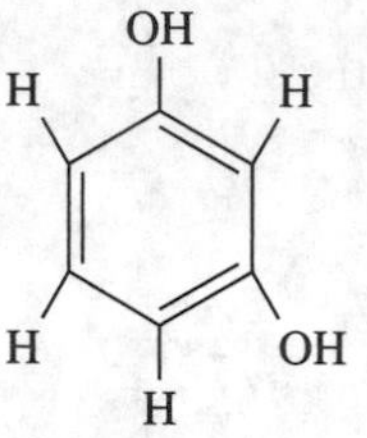

5-9
1. tertiary
2. secondary
3. primary
4. tertiary
5. secondary

5-10
1. *tert*-hexyl alcohol or 2-methyl-2-pentanol
2. 2,4-hexandiol
3. *n*-propanol
4. 2-methyl-2-pentanol
5. 1,2-cyclohexandiol

5-11
1. $C_5H_{11}OH$, 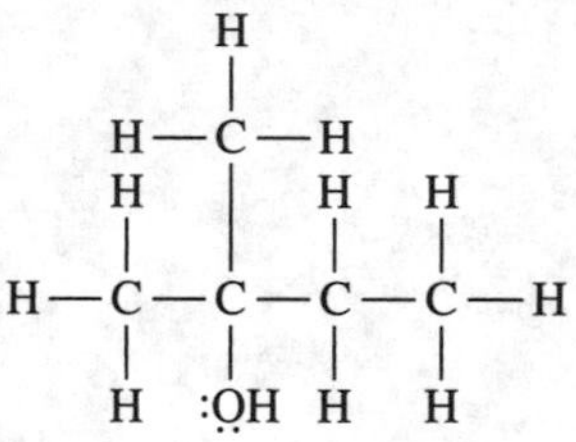 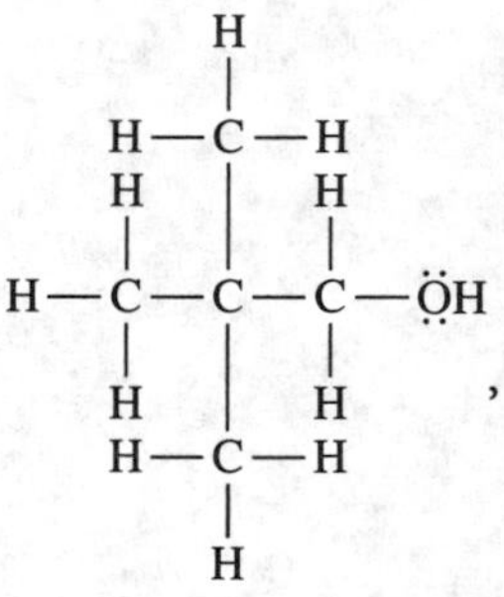 , 2-methyl-2-butanol, 2,2-dimethyl-propanol

2. $C_6H_{13}OH$, 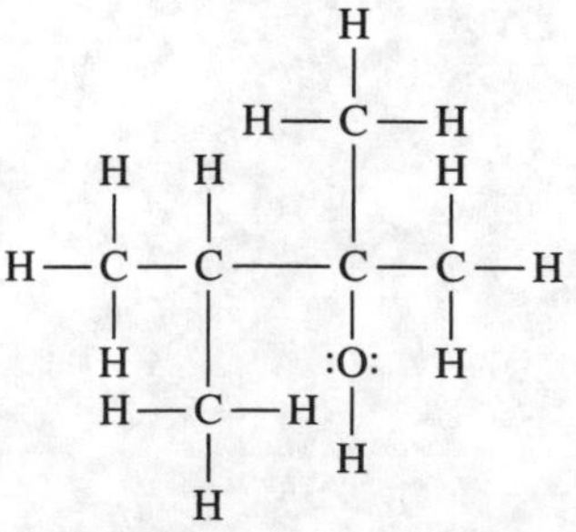 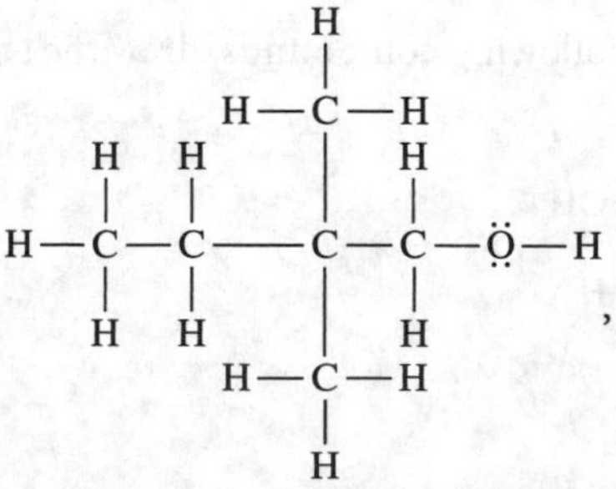 , 2,3-dimethyl-2-butanol, 2,2-dimethyl-1-butanol

3. $C_7H_{15}OH$, 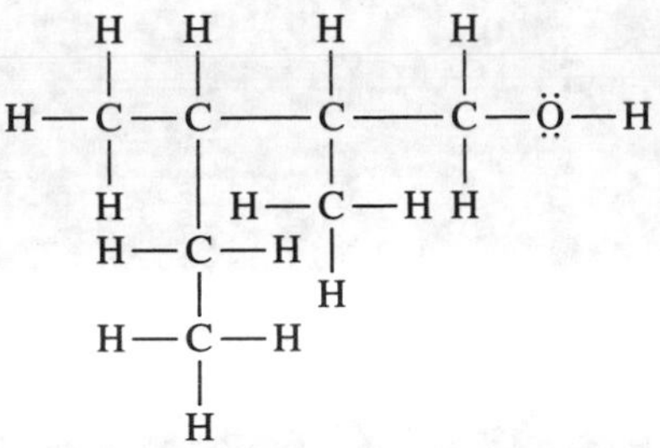 , 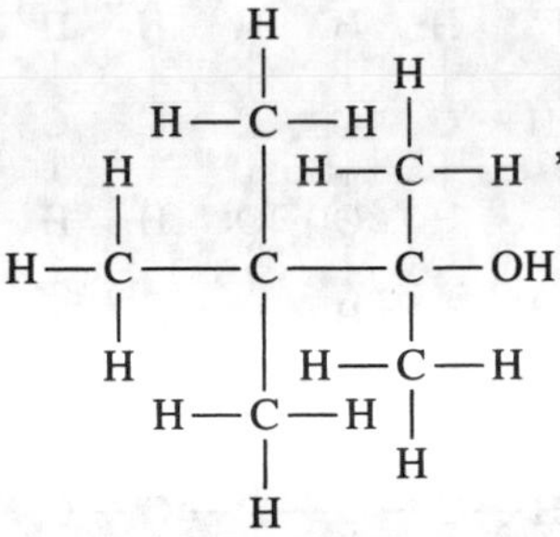 , 2,3-dimethyl – 1-pentanol, 1,1,2,2-tetramethyl-1-propanol

4. $C_8H_{17}OH$, 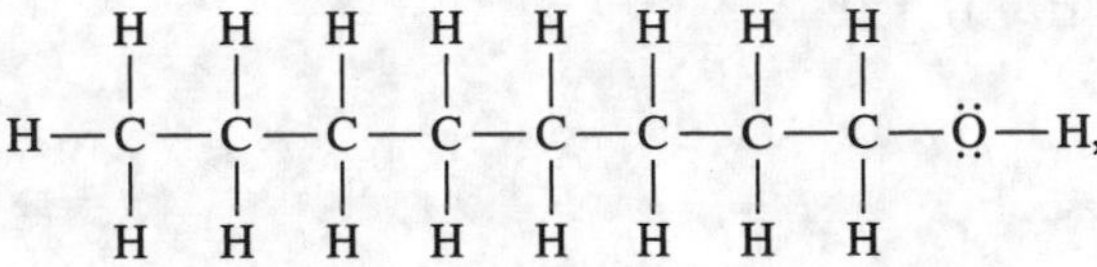, *n*-octanol, 4-ethyl-1-hexanol

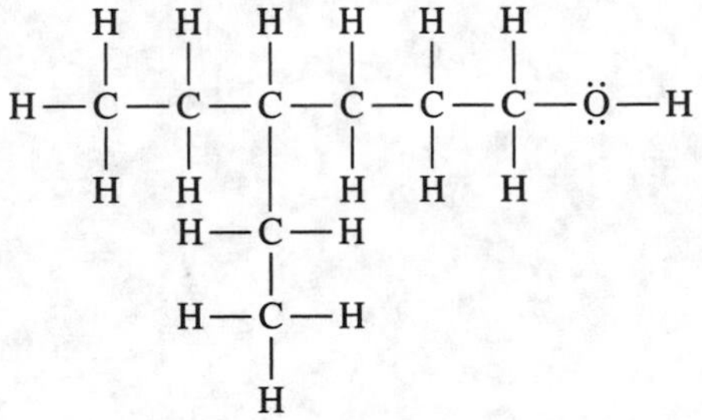

5. $C_{10}H_{21}OH$, 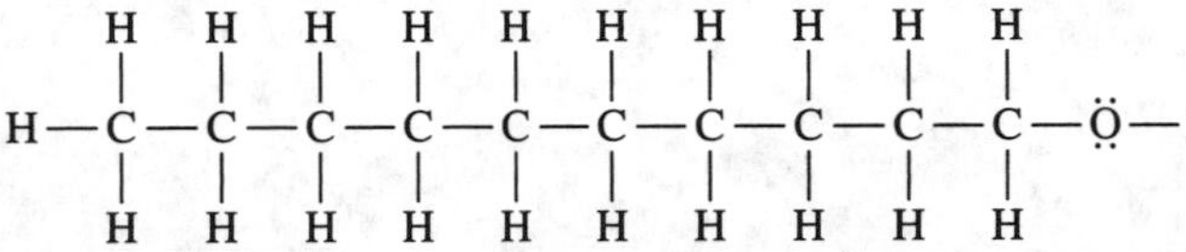, *n*-decanol, 4-ethyl-5-methyl-2-heptanol

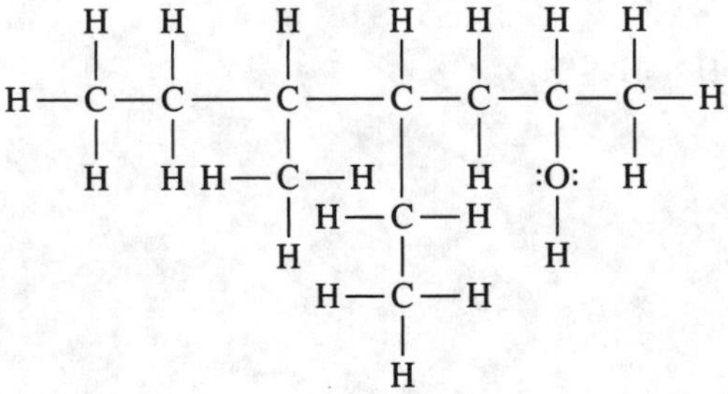

5-12
1. circle propanol
2. circle propandiol
3. circle butanol
4. circle propantriol
5. circle butandiol

5-13
1. 6-methyl-1-heptanol, $C_8H_{17}OH$
2. cyclopropanol, C_3H_5OH
3. 2,3-pentandiol, $C_5H_{10}(OH)_2$
4. 2-methyl-2-pentanol, $C_6H_{13}OH$

5-14
1. CH_3OH, 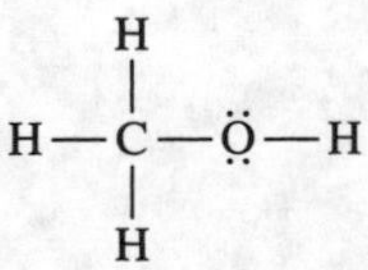

2. $C_5H_{10}(OH)_2$,

```
      H   H   H   H   H
      |   |   |   |   |
  H — C — C — C — C — C — H
      |   |   |   |   |
      H  :O: :O:  H   H
          |   |
          H   H
```

3. $C_6H_{11}(OH)_3$,

```
      H  OH  H   H   H   H
      |   |   |   |   |   |
  H — C — C — C — C — C — C — H
      |   |   |   |   |   |
      H  OH  OH  H   H   H
```

4. C_3H_7OH,

```
      H   H   H
      |   |   |
  H — C — C — C — H
      |   |   |
      H  :OH  H
```

5-15   1. 2°   2. 1°   3. 3°   4. 2°

5-16   1.

```
          H   H
          |   |
  H — Ö — C — C — Ö — H
          |   |
          H   H
```

2.

```
          H   H   H
          |   |   |
  H — Ö — C — C — C — Ö — H
          |   |   |
          H  :O:  H
              |
              H
```

3.

```
      H   H   H   H   H   H   H   H
      |   |   |   |   |   |   |   |
  H — C — C — C — C — C — C — C — C — Ö — H
      |   |   |   |   |   |   |   |
      H   H   H   H   H   H   H   H
```

4.

```
      H   H     H       H
      |   |     |       |
  H — C — C —   C   —   C — H
      |   |     |       |
      H  :Ö:    H       H
          |
          H
```

5-17   1.

```
      H
       \
        C — Ö — H
       /|
      H H
```

2.

```
          H
          |
      H   C      Ö — H
       \  ‖     /
        C — C
       /|   |\
      H H   H
```

5-18 1. $CH_3C(CH_3)OHCH_2CH_2CH_3$
2. $CH_3CHOHCH_2CH_2CH_2OH$
3. $CH_3CHOH(CH_2)_5CH_3$
4. $CH_2OHCHOHCH_2OH$

5-19 1.

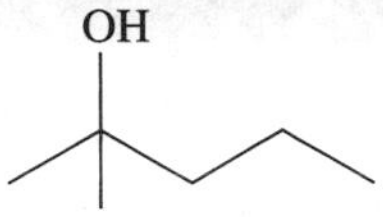

2.

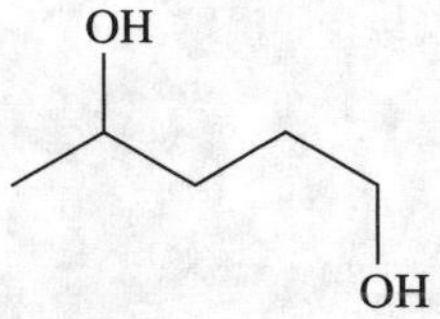

3.

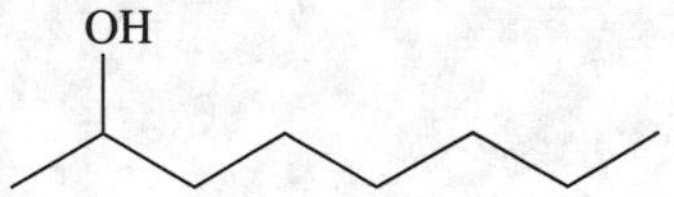

4.

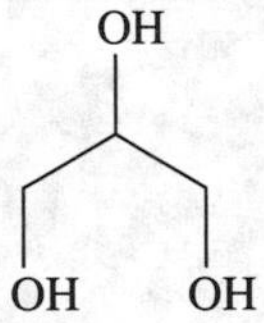

5-20 1. Isomer 1

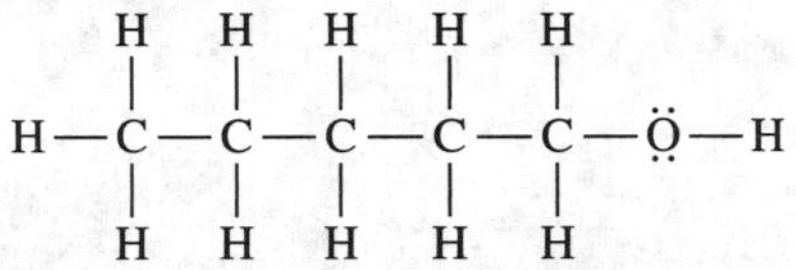

n-pentanol
$CH_3CH_2CH_2CH_2CH_2OH$

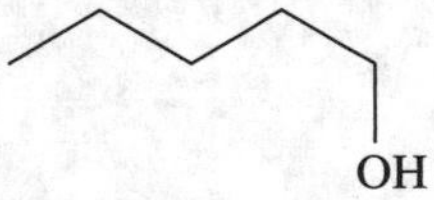

Isomer 2

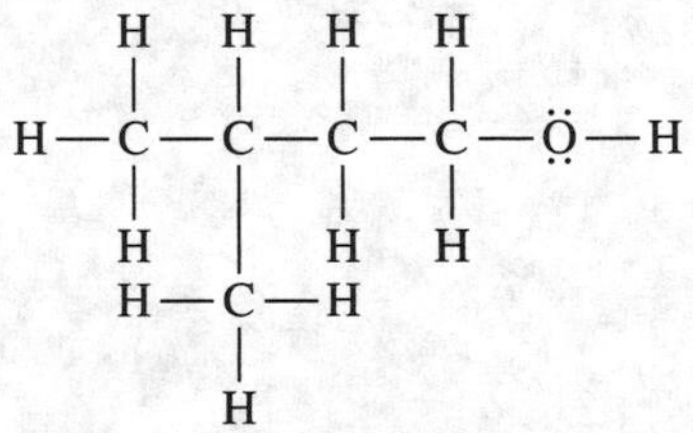

3-methyl-1-butanol
$CH_3CH(CH_3)CH_2CH_2OH$

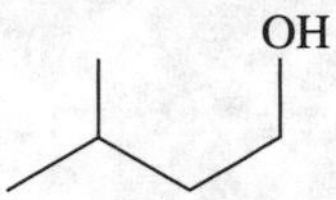

2. Isomer 1

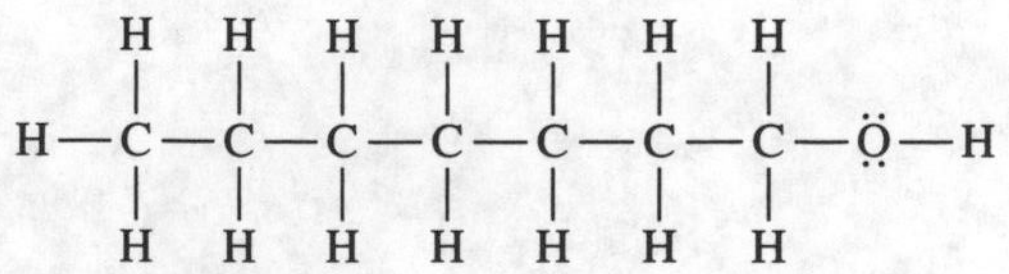

H-C-C-C-C-C-C-C-Ö-H

n-heptanol

$CH_3CH_2CH_2CH_2CH_2CH_2CH_2OH$

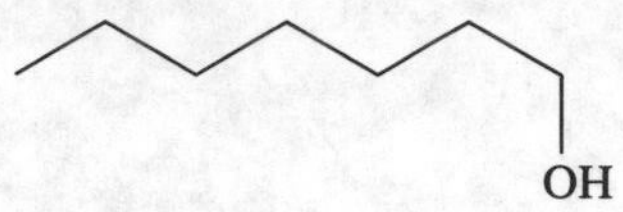

OH

Isomer 2

H-C-C-C-C-C-Ö-H

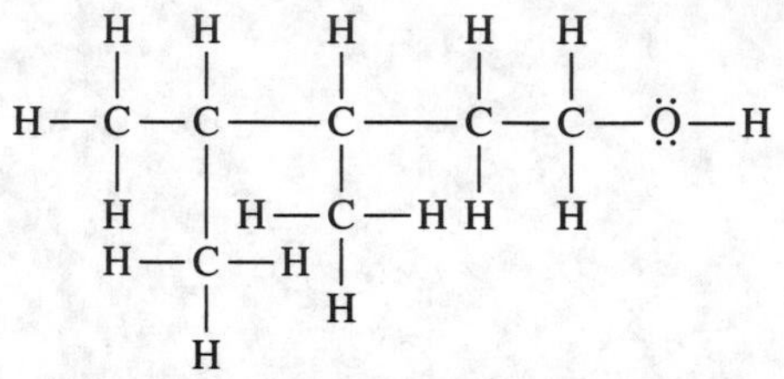

3,4-dimethyl-1-pentanol

$CH_3CH(CH_3)CH(CH_3)CH_2CH_2OH$

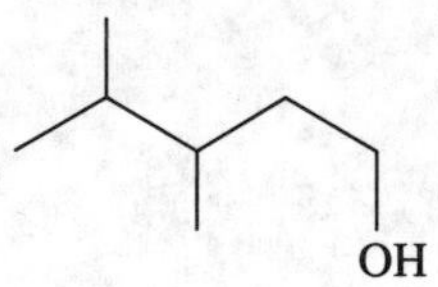

OH

3. Isomer 1

H-C-C-C-C-C-C-Ö-H

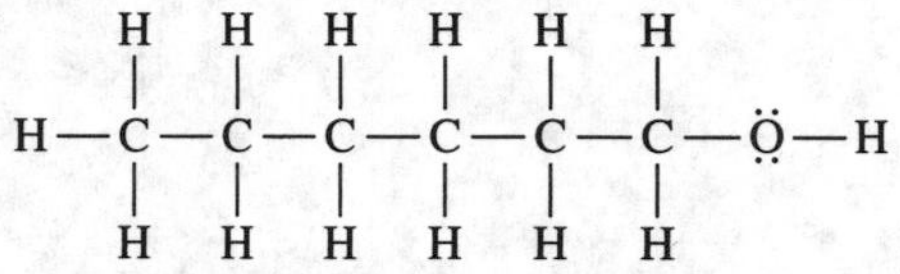

n-hexanol

$CH_3CH_2CH_2CH_2CH_2CH_2OH$

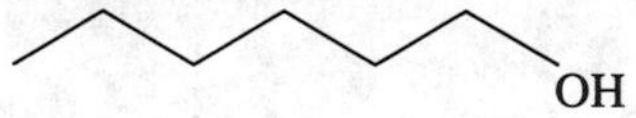

OH

Isomer 2

$H_3C-C-CH_2CH_2CH_2OH$

CH_3

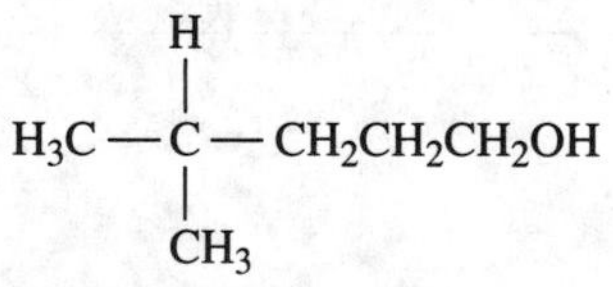

3-methyl-1-pentanol

$CH_3CH_2CH(CH_3)CH_2CH_2OH$

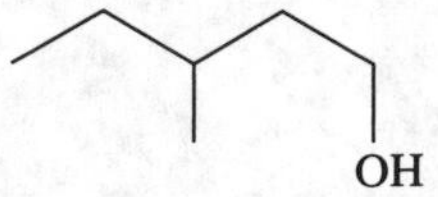

OH

4. Isomer 1

H—C—C—C—C—Ö—H

(with H atoms above and below each of the four carbons)

n-butanol
$CH_3CH_2CH_2CH_2OH$

Isomer 2

H—C—C—C—Ö—H

(with a CH_3 branch: H—C—H below the middle carbon)

2-methyl-1-propanol
$CH_3CH(CH_3)CH_2OH$

5-21   1. C_3H_7OH, 2-propanol, $CH_3CHOHCH_3$

2. $C_4H_8(OH)_2$, 2,3-butandiol, $CH_3CHOHCHOHCH_3$

3. $C_6H_{13}OH$, *n*-hexanol, $CH_3CH_2CH_2CH_2CH_2CH_2OH$

4. $C_7H_{15}OH$, 3-methyl-3-hexanol, $CH_3CH_2C(CH_3)OHCH_2CH_2CH_3$

5-22   1. Both have hydrogen bonding, but propanol has greater dispersion forces and a higher dipole-dipole force.

2. Propanol has hydrogen bonding that propane does not have. This increased force increases the amount of energy required to pull molecules away from each other to enter the gas phase.

3. 1,2,3-Propanetriol has a higher boiling point as a consequence of increased amounts of hydrogen bonding.

6   Aldehydes

6-1   1. hexanal, $C_6H_{12}O$

2. 2,4-dibromohexanal, $C_6H_{10}OBr_2$

3. 4-bromo-2,5-dimethylhexanal, $C_8H_{15}OBr$

4. cyclopentanecarbaldehyde, $C_6H_{10}O$

5. 2-butenal, C_4H_6O

6-2   1. H—C—C—C—C—C—H, C_5H_9OCl

(with H atoms and a Cl substituent; double-bonded O on terminal carbon)

2. H—C—C—C—C—H, $C_6H_{12}O$

(with CH_2CH_3 branch; double-bonded O on terminal carbon)

3. 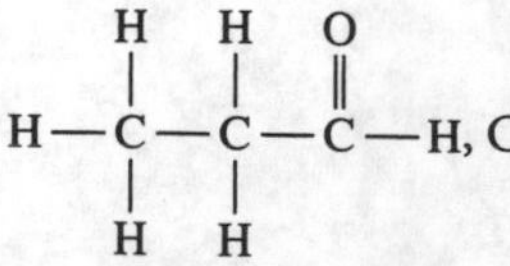 C_3H_6O

4. 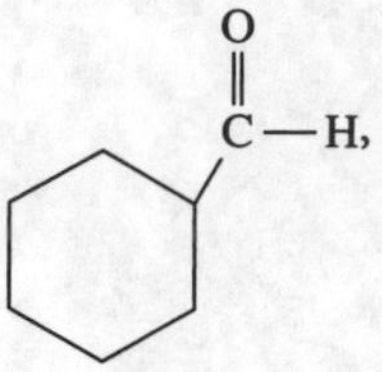 $C_6H_{12}O$

5. 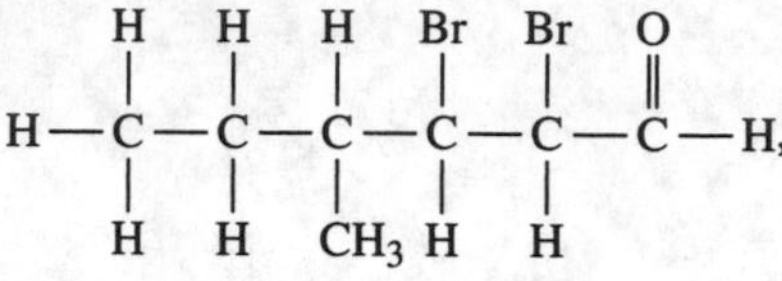 $C_7H_{12}OBr_2$

6-3 1. 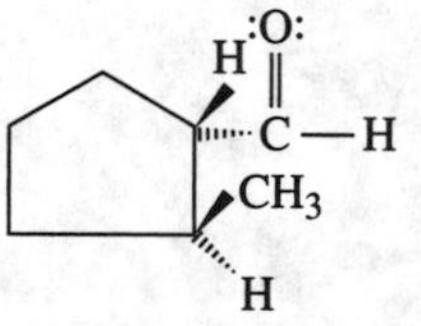

2. 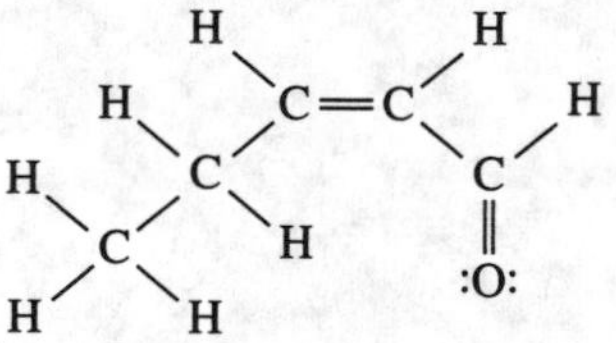

3. 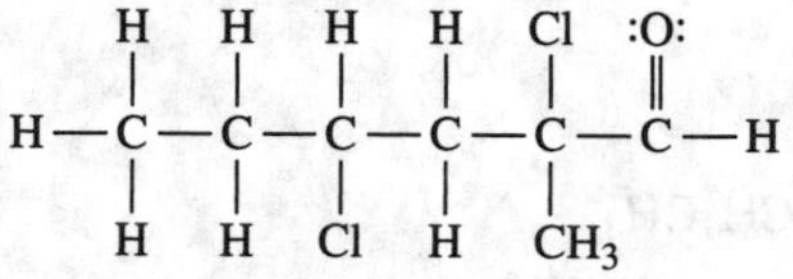

4. 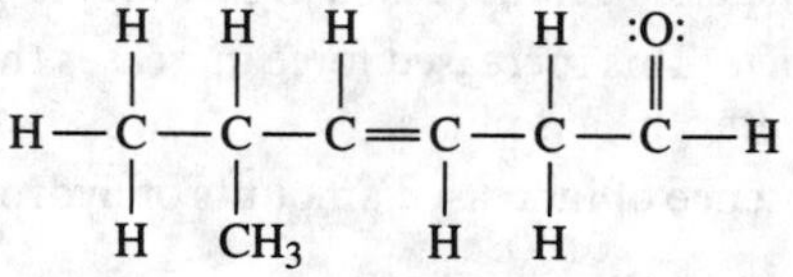

5. 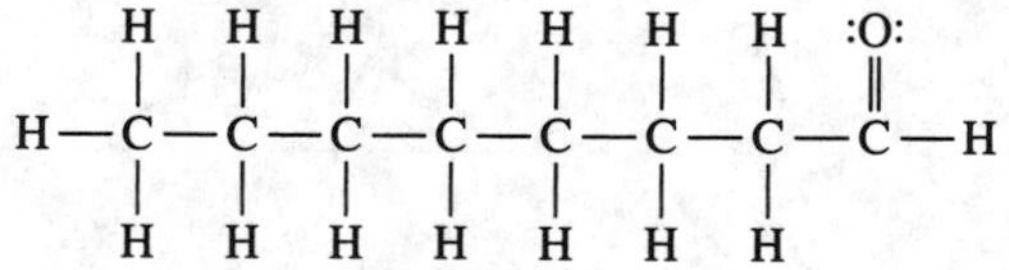

6-4 1. 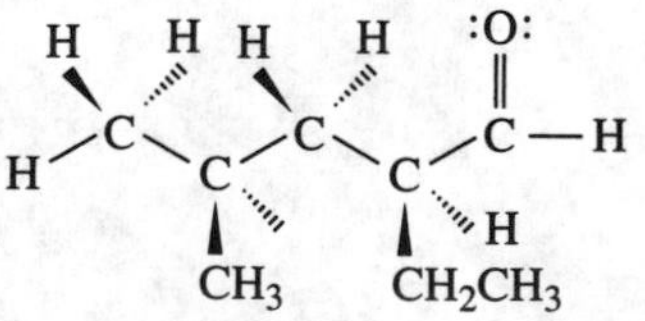

2. 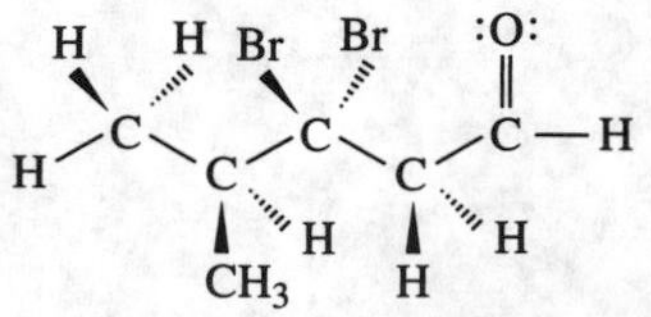

3.

4.

5.

6-5 For

1.

2.

3.

4.

5.

6-6 1.

2.

3.

4.

5.

6-7 Answers will vary.

1. CH_3C—C ... 2-bromo-propanal

2. $CH_3CH_2CH_2CH_2CH_2C$—H hexanal

3. $CH_3(CH_2)_7C$—H nonanal

4. $CH_3CH_2CHCH_2CHCH_2C_6H_2C$ ... 4,6-dichloro-octanal

6-8 1. $C_6C_{12}O$ 2,3-dimethyl-butanal

2. $C_6H_9OBr_3$, 3,4,5-tribromo-hexanal, $CH_3CHCHCHCH_2C$—H

3. $(C_6H_5)C_4H_7O$, 4-phenylbutanal,

4. $C_8H_{14}OCl_2$, 2,4-dichloro-3,5-dimethyl-hexanal,

$$
\begin{array}{ccc}
CH_3 & CH_3 & :O: \\
| & | & \| \\
CH_3CHCHCHCHC & & -H \\
| & | & \\
Cl & Cl &
\end{array}
$$

5. $C_8H_{14}O$, 4-methyl-heptanedial,

$$
\begin{array}{ccc}
:O: & CH_3 & :O: \\
\| & | & \| \\
H-CCH_2CH_2CHCH_2CH_2C &
\end{array}
$$

6-9

1. Butanal and butanol both have dispersion forces, dipole-diploe forces, and hydrogen bonding. Butanol will have the higher boiling point. Even though they have the same forces, these forces are stronger in butanol.

2. Pentanal and pentanol both have dispersion forces, dipole-diploe forces, and hydrogen bonding. Pentanol will have the higher boiling point. Even though they have the same forces, these forces are stronger in pentanol.

3. Hexanal and hexanol both have dispersion forces, dipole-diploe forces, and hydrogen bonding. Hexanol will have the higher boiling point. Even though they have the same forces, these forces are stronger in hexanol.

4. Propanal and propane both have dispersion forces, but propanal also has dipole-dipole forces and hydrogen bonding. Propanal will have the higher boiling point.

5. Butanal and 2-butene both have dispersion forces, but butanal also has dipole-dipole forces and hydrogen bonding. Butanal will have the higher boiling point.

7   Ketones

7-1

1. 4-methyl-3-hexanone, $C_6H_{14}O$

2. 3-nonanone, $C_9H_{18}O$

3. 3,5-dimethyl-4-heptanone, $C_9H_{18}O$

4. 7-methyl-2-decanone, $C_{11}H_{22}O$

5. 5,6,6-trimethyl-3-heptanone, $C_{10}H_{20}O$

7-2

1. C_5H_9OCl

2. $C_6H_{12}O$

3. C_3H_5OBr

4. $C_6H_{10}O$

5. 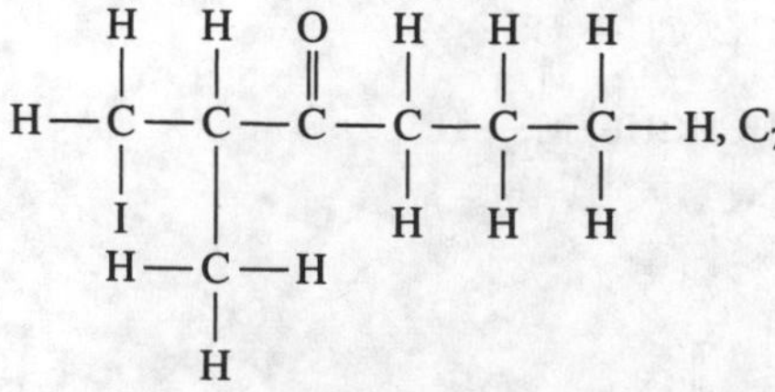 , $C_7H_{13}OI$

7-3

1. 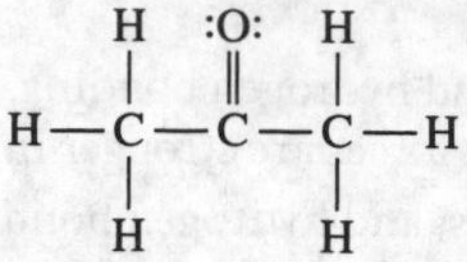

2. 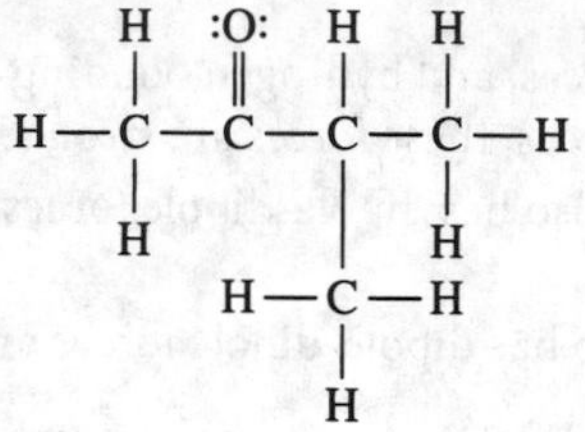

3. 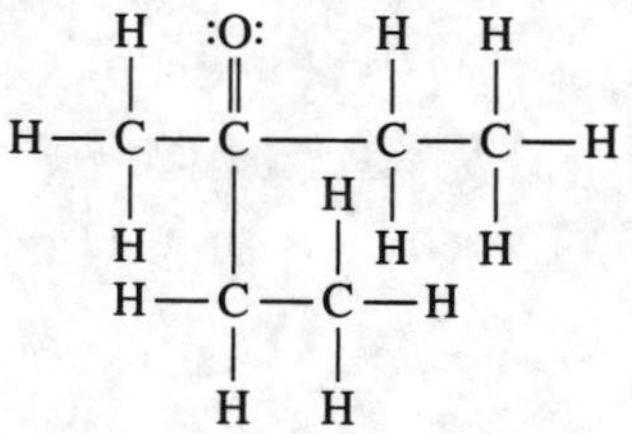

4. 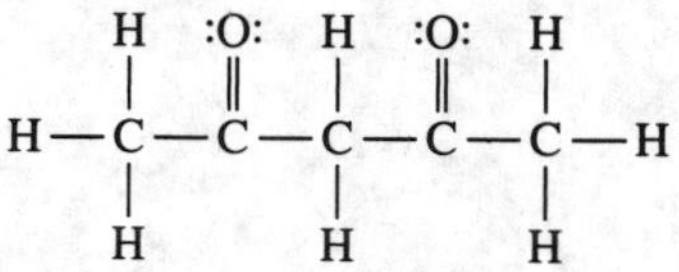

5. 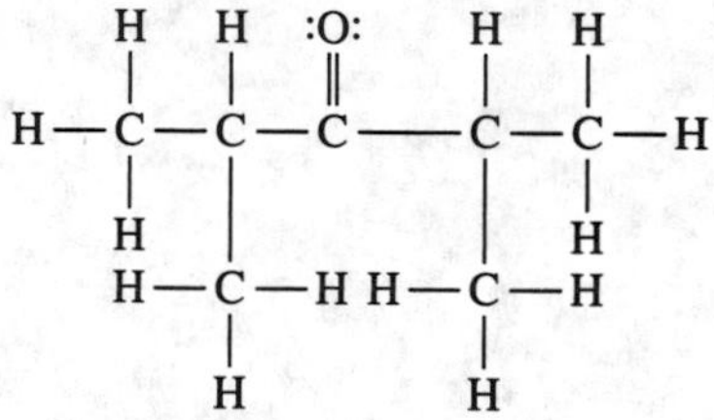

7-4

1. 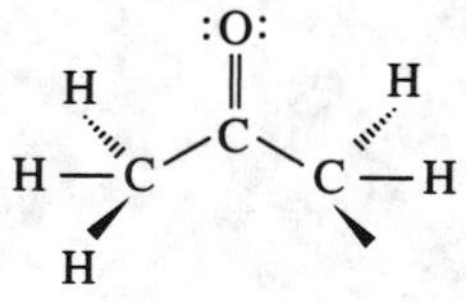

2. 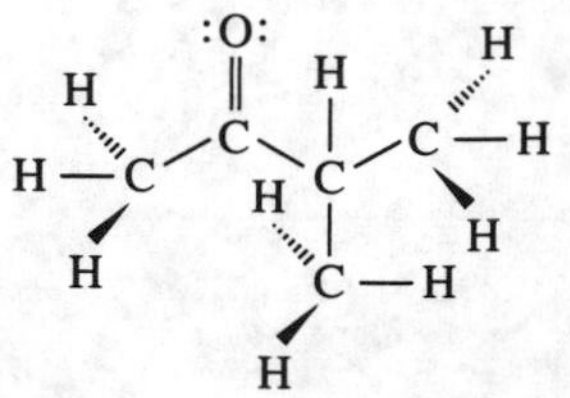

3.

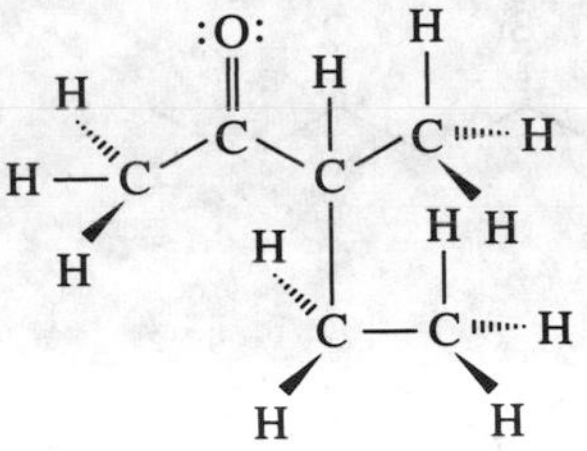

4.

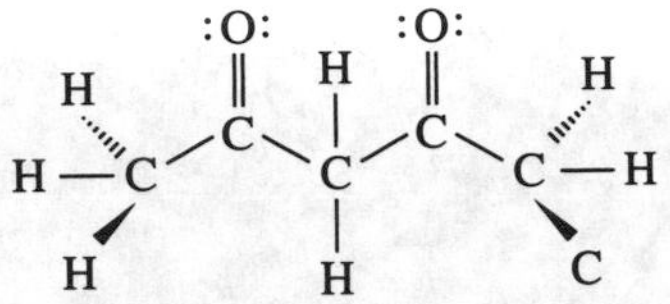

5.

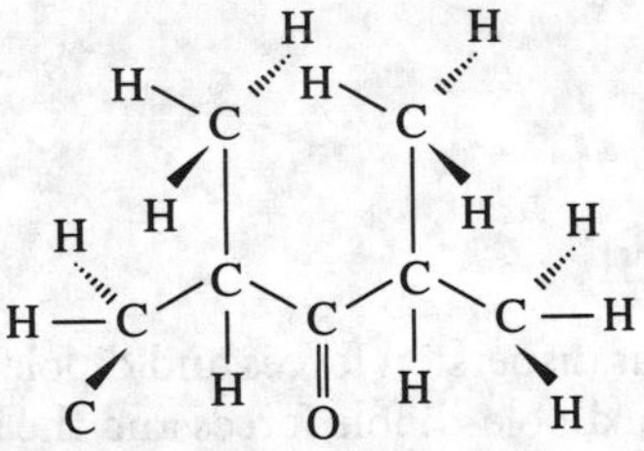

7-5 1. CH_3COCH_3

2. $CH_3COCH(CH_3)CH_3$

3. $CH_3COCH(CH_2CH_3)CH_2CH_3$

4. $CH_3COCH_2COCH_3$

5. $CH_3CH(CH_3)COCH(CH_3)CH_3$

7-6 1.

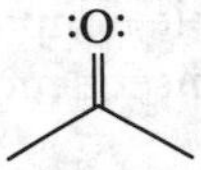

2.

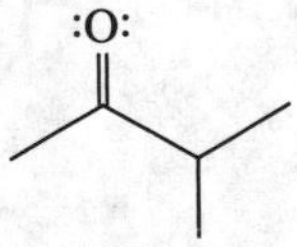

3.

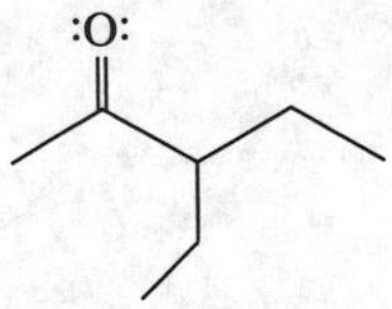

4.

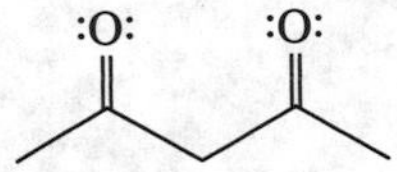

5.

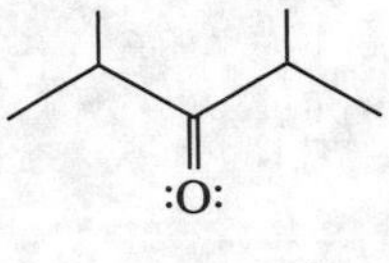

7-7 1. $CH_3CH_2CH_2 - \overset{\overset{\displaystyle :O:}{\|}}{C} - CH_2CH_2CH_3$ 4-heptanone

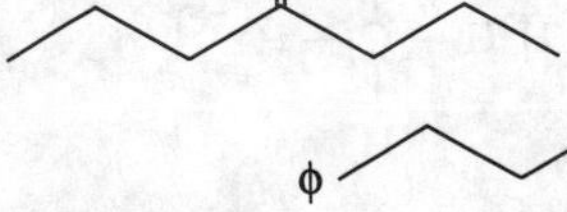

2.

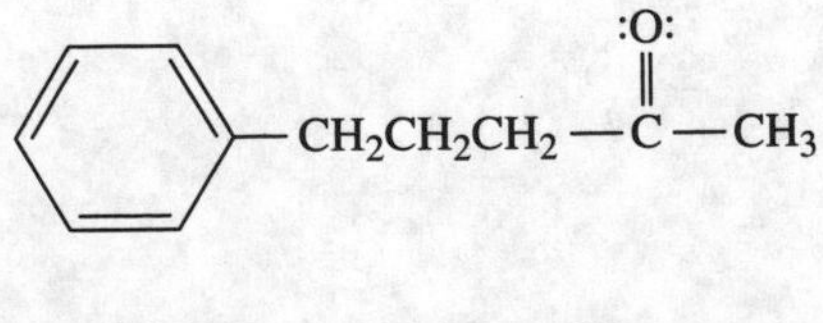

$- CH_2CH_2CH_2 - \overset{\overset{\displaystyle :O:}{\|}}{C} - CH_3$ 5-phenyl-2-pentanone

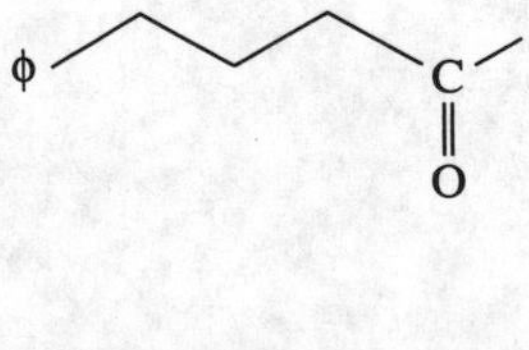

3.

$$H-\underset{\underset{H}{|}}{\overset{\overset{Br}{|}}{C}}-\underset{\underset{Br}{|}}{\overset{\overset{H}{|}}{C}}-\underset{\underset{H}{|}}{\overset{\overset{H}{|}}{C}}-\overset{\overset{:O:}{||}}{C}-CH_2CH_3$$

5, 6-dibromo-3-hexanone

4.

$$\phi-CH_2-\overset{\overset{:O:}{||}}{C}-CH_2-\underset{\underset{Cl}{|}}{\overset{\overset{Cl}{|}}{C}}-H$$

4,4-dichloro-
1-phenyl-2-
butanone

7-8

1. $C_7H_{12}O$, 6-heptene-2-one, $CH_2CHCH_2CH_2CH_2COCH_3$

2. $C_6H_{10}O_2$, 2,4-hexandione, $CH_3CH_2COCH_2COCH_3$

3. $C_9H_{10}O$, 1-phenyl-2-butanone, $(C_6H_5)COCH_2CH_3$

4. $C_7H_{14}O$, 2-heptanone, $CH_3COCH_2CH_2CH_2CH_3$

5. $C_{10}H_{20}O$, 3-decanone, $CH_3CH_2COCH_2CH_2CH_2CH_2CH_2CH_3$

7-9

1. Nonpolar propane has dispersion forces, whereas propanone has dispersion forces and dipole-dipole forces. Propanone will have the higher boiling point because of the dipole-dipole forces and the increased dispersion forces.

2. Nonpolar 2-pentene has dispersion forces, whereas 2-pentanone has dispersion forces and dipole-dipole forces. With increased intermolecular forces, 2-pentanone has the higher boiling point.

3. 2-Butanone and 3-hexanone have similar forces attracting molecules to each other. With the increased dispersion forces that result from its having more electrons, 3-hexanone will have the higher boiling point.

4. Ethyne is nonpolar and has dispersion forces, whereas 2-pentanone has both dispersion forces and dipole-dipole interactions, resulting in 2-pentanone having a higher boiling point.

5. 2,3-Butanedione has dispersion forces and dipole-dipole interactions, whereas while nonpolar butane has only dispersion forces. 2,3-Butanedione will have the higher boiling point as a consequence of the increased forces.

8 Carboxylic Acids

8-1

1. hepatandioic acid

2. benzedioic acid

3. β-Br-nonandioic acid

4. α-Br-hexandioic acid

5. α Cl-β-Br-γ-Cl-pentandioic acid

8-2

1.

$$H-\underset{\underset{H}{|}}{\overset{\overset{H}{|}}{C}}-\underset{\underset{H}{|}}{\overset{\overset{H}{|}}{C}}-\underset{\underset{:Br:}{|}}{\overset{\overset{H}{|}}{C}}-\underset{\underset{:Cl:}{|}}{\overset{\overset{H}{|}}{C}}-\overset{\overset{:O:}{||}}{\underset{\underset{:O:}{\diagdown}}{C}}H$$

2.

$$H-\underset{\underset{H}{|}}{\overset{\overset{H}{|}}{C}}-\underset{\underset{H}{|}}{\overset{\overset{H}{|}}{C}}-\underset{\underset{H}{|}}{\overset{\overset{:\ddot{O}-H}{|}}{C}}-\underset{\underset{H-\underset{\underset{H}{|}}{\overset{\overset{H}{|}}{C}}-H}{|}}{\overset{\overset{H}{|}}{C}}-\overset{\overset{:O:}{||}}{\underset{\underset{:O:}{\diagdown}}{C}}H$$

3. 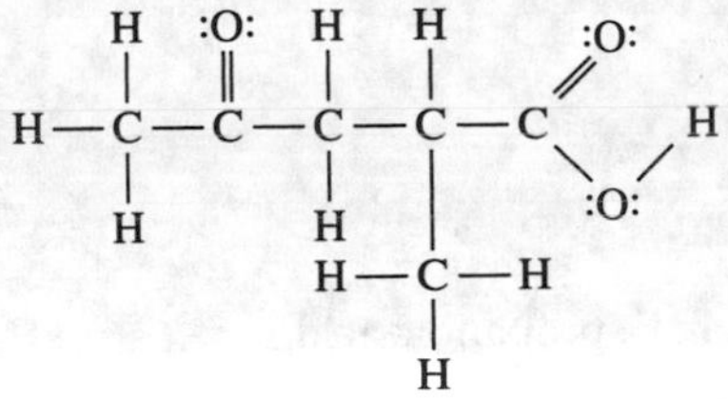

4. 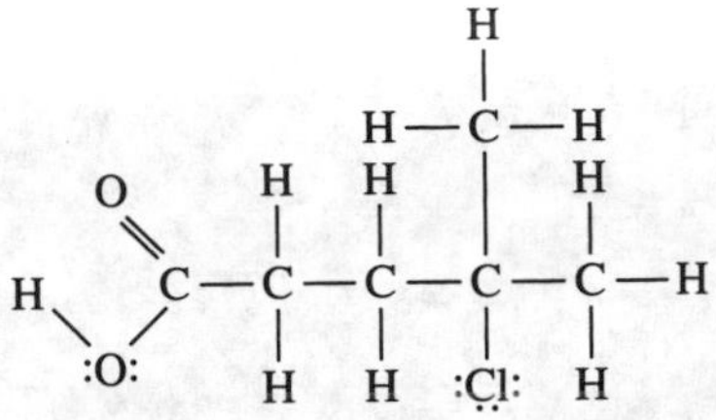

5. 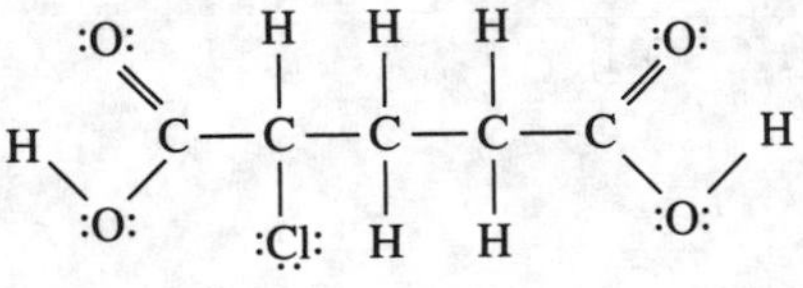

8-3 1. $CH_3CHClCH_2CH_2CH_2COOH$, 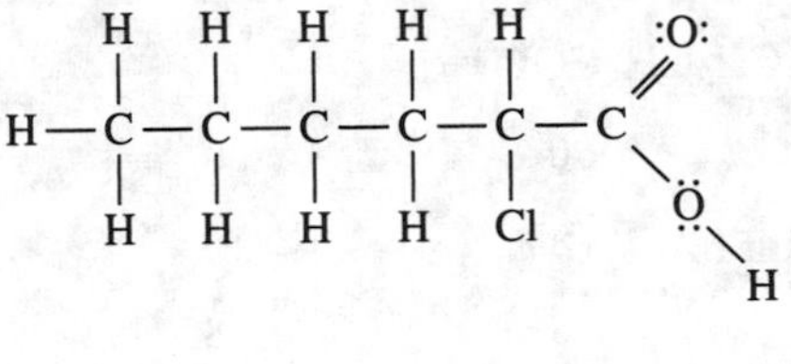 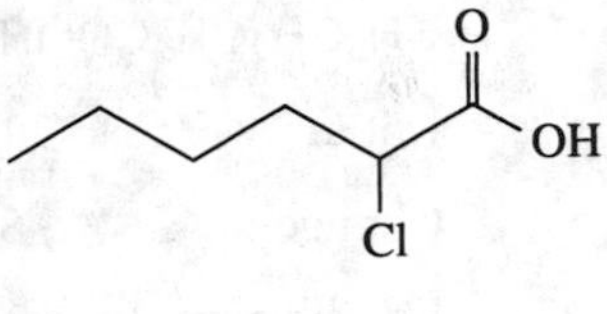

2. HOOC-CH_2-COOH, 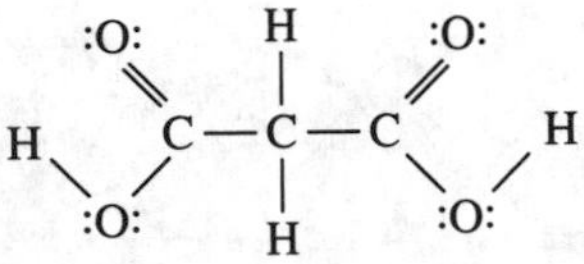 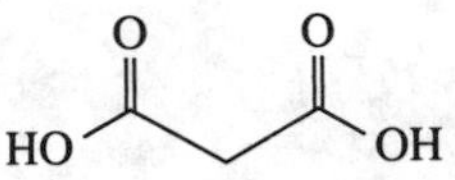

3. $CH_3CH_2CH(CH_3)CH_2CHOHCH_2COOH$,

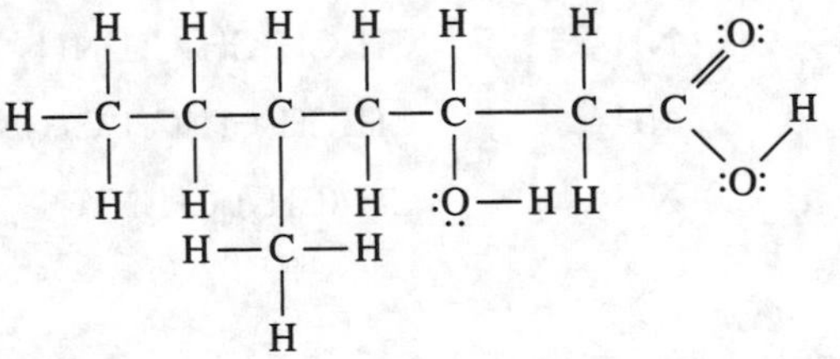 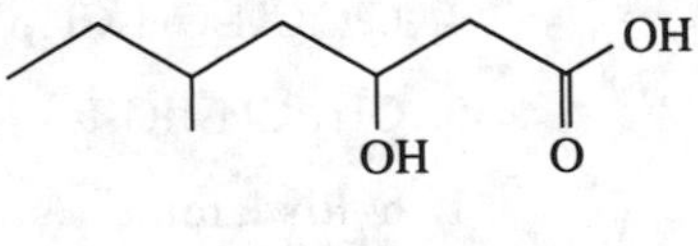

4. $HOOCCH_2CHClCOOH$, 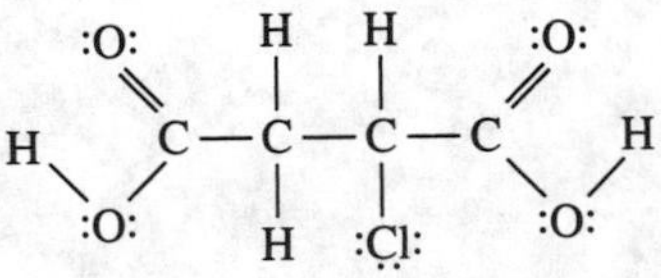 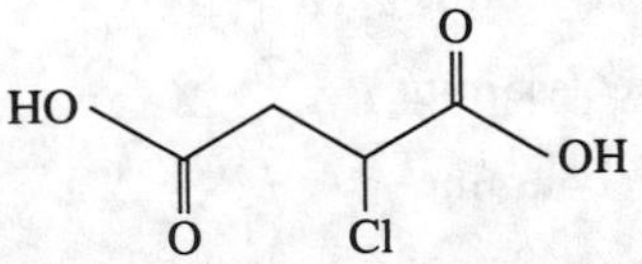

8-4 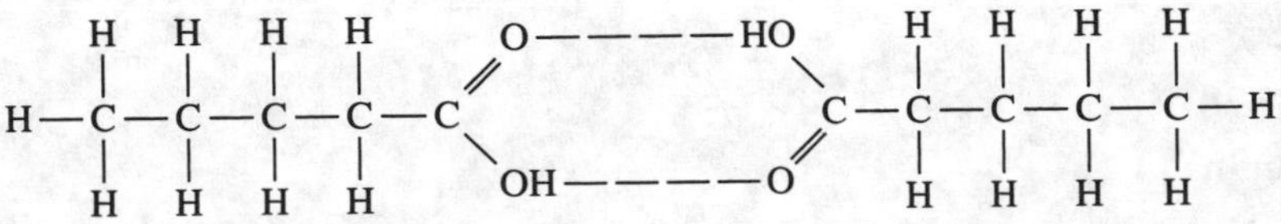

The boiling point will be greater than 141°C (the boiling point of propanoic acid).

8-5 Dimers have hydrogen bonds between molecules, which require more energy to pull molecules apart from each other and into the gas phase.

8-6

$$H-\overset{\displaystyle H}{\underset{\displaystyle H}{C}}-\overset{\displaystyle H}{\underset{\displaystyle H}{C}}-\overset{\displaystyle H}{\underset{\displaystyle H}{C}}-\overset{\displaystyle H}{\underset{\displaystyle H}{C}}-\overset{\displaystyle :O:}{\underset{\displaystyle :O:}{C}}-H$$

8-7 Ethanoic acid has a larger K, so it will have a greater amount of H^+. propanoic acid.

$$CH_3COOH \rightleftharpoons CH_3COO^-(aq) + H^+(aq)$$

$$K_a = \frac{[CH_3COO^-][H^+]}{[CH_3COOH]}$$

8-8 0.0013 M or 1.3×10^{-3} M

8-9 2.89

8-10 Methanoic acid, $HCOOH \rightleftharpoons HCOO^-(aq) + H^+(aq)$, $K = \dfrac{[HCOO^-][H^+]}{[HCOOH]}$

8-11 0.0042 M or 4.2×10^{-3} M

8-12 2.37

8-13 192 remain because only 8 dissociate. $CH_3CH_2CH_2COOH$ (aq) $\rightleftharpoons CH_3CH_2CH_2COO^-(aq) + H^+(aq)$, $200 - (200 \times 0.04) = 192$.

	$CH_3CH_2CH_2COOH$	$CH_3CH_2CH_2COO^-$	H^+
Initial	200	0	0
Change	−8	+8	+8
Equilibrium	192	8	8

8-14 pH = 2.6, 4%

9   Amines

9-1
1. CH_3CH_2 NH_2 $CH_3CH_2NH_2$
2. $CH_3CH_2CH_3$ NH_2 $CH_3CH_2CH_2NH_2$
3. $CH_3CH_2CH_2CH_3$ NH_2 $CH_3CH_2CH_2CH_2NH_2$
4. $CH_3CH_2CH_2CH_2CH_3$ NH_2 $CH_3CH_2CH_2CH_2CH_2NH_2$
5. $CH_3(CH_2)_6CH_3$ NH_2 $CH_3(CH_2)_6CH_2NH_2$

9-2
1. n-ethylamine
2. n-propylamine
3. n-butylamine
4. n-pentylamine
5. n-octylamine

9-3
1. 2-bromo-1-propylamine
2. 1-bromo-2-propylamine
3. 3-methyl-1-pentalamine
4. 2,4-dimethyl-2-pentylamine
5. 2,4,4-trimethyl-3-pentylamine

9.4 1. *tert*-butylamine

2. ethyl methylamine

3. N-methyl-2-methyl-3-pentanamine

4. N,N-diethyl-1-butylamine

5. N,N-dimethyl-1-propylamine

9.5 1.

2.

$\text{C}_6\text{H}_5-\text{CH}_2\text{CH}_2\text{NH}_2$

3.

$$\text{CH}_3\text{CHCH}_2\text{CHCH}_2\text{CH}_3$$
$$\text{CH}_3\text{CH}_2-\overset{\cdot\cdot}{\text{N}}-\text{H} \quad \text{CH}_3$$

4.

$$\text{CH}_2\text{CH}_3$$
$$:\text{N}-\text{CH}_2\text{CH}_3$$
$$\text{CH}_2\text{CH}_3$$

5. $\text{CH}_3\text{CH}_2\text{CH}_2\text{CH}_2\text{CH}_2\text{CH}_2-\overset{\cdot\cdot}{\underset{\text{H}}{\text{N}}}-\text{CH}_2\text{CH}_2\text{CH}_3$

9.6 1. 3-amino-1-propanol

2. 3-aminopropanoic acid or β-Alanine

3. 2-(N-methylamino)propanol

4. *p*-benzenediamine

5. 4-amino-2-butanone

9.7 1.

$$\text{CH}_2\text{CH}_2\overset{\cdot\cdot}{\text{O}}\text{H}$$
$$:\text{NH}_2$$

2.

3.

$$:\text{O}:$$
$$\text{CH}_3\text{CHCH}_2\text{CH}_2\overset{||}{\text{C}}-\overset{\cdot\cdot}{\text{O}}\text{H}$$
$$:\text{NH}_2$$

4.

$$:\text{O}:$$
$$\text{CH}_2\text{CH}_2\text{CH}_2\text{CH}_2\overset{||}{\text{C}}-\text{H}$$
$$:\text{NH}_2$$

5.

$$:\text{O}:$$
$$\text{CH}_2\text{CH}_2\overset{||}{\text{C}}-\text{CH}_2\text{CH}_3$$
$$:\text{NH}_2$$

9.8
1. N-propylcyclohexylamine
2. *m*-chloroaniline
3. *o*-nitroaniline
4. N-methyl-2-phenylethylamine
5. N-ethyl-N-methylaniline

9.9

1. 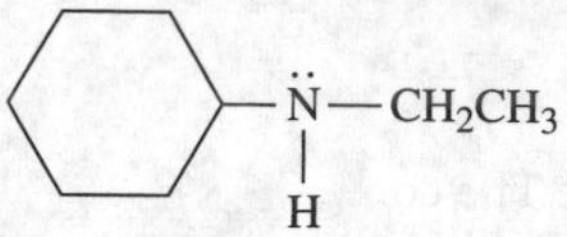 cyclohexyl—N̈(H)—CH₂CH₃

2. 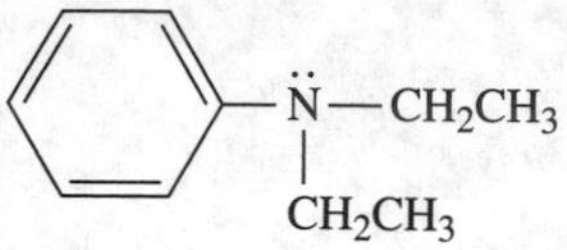 phenyl—N̈(CH₂CH₃)—CH₂CH₃

3. 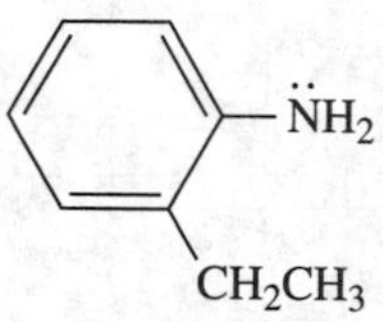 phenyl with N̈H₂ and CH₂CH₃ (ortho)

4. 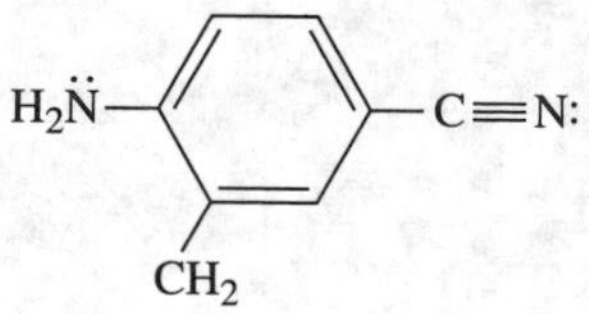 H₂N̈—(ring)—C≡N: with CH₂

5. 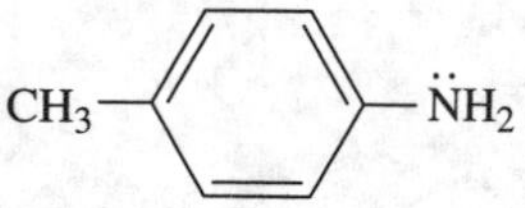 CH₃—(ring)—N̈H₂

9.10
1. primery amine
2. tertiary amine
3. secondary amine
4. secondary amine
5. tertiary amine

9.11 All compounds have the same molecular formula, C_3H_9NO, and have the same molecular weight.

lowest building point compound

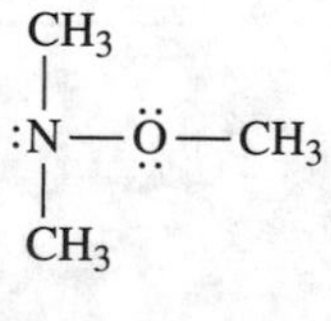

$$:N(CH_3)(CH_3)—\ddot{O}—CH_3$$

no hydrogen bonding IMF between molecules. Two IMFs present are dipole-dipole and London dispersion forces.

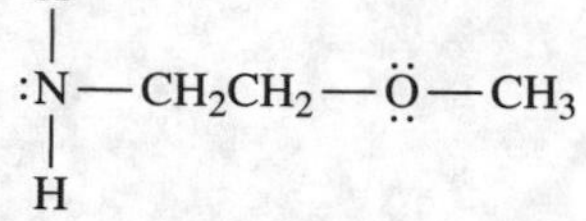

$$:N(H)(H)—CH_2CH_2—\ddot{O}—CH_3$$

The $\overset{\delta-}{:N-}\overset{\delta+}{H}$ is a polar covalent bond. The partial negative change on the nitrogen and the partial positive charge on the hydrogen are capable of forming a hydrogen bond between molecules. Dipole-dipole and Londan dispersion IMFs are also present.

Highest boiling point compound

$$\overset{\displaystyle H}{\underset{\displaystyle H}{\overset{\displaystyle |}{\underset{\displaystyle |}{:N}}}}-CH_2CH_2CH_2-\overset{..}{O}-H$$

In addition to the NH_2 group, this compound has the O—H group. The compound has two areas capable of forming hydrogen bonds with other molecules. Dipole-dipole and London dispersion IMFs are also present.

9.12 1. $CH_3CH_2\overset{..}{\underset{\displaystyle H}{\overset{\displaystyle |}{N}}}-H + HCl \longrightarrow CH_3CH_2\overset{\oplus}{\underset{\displaystyle H}{\overset{\displaystyle |}{N}}}-H, Cl^{\ominus}$

2. $CH_3CH_2\overset{..}{\underset{\displaystyle H}{\overset{\displaystyle |}{N}}}-CH_3 + H\overset{..}{O}\overset{\displaystyle :O:}{\overset{\displaystyle \|}{C}}-CH_3 \rightleftharpoons CH_3CH_2-\overset{\oplus}{\underset{\displaystyle H}{\overset{\displaystyle H}{\overset{\displaystyle |}{N}}}}-CH_3, \ :\overset{..}{\underset{..}{O}}-\overset{\displaystyle :O:}{\overset{\displaystyle \|}{C}}-CH_3$

9.13 ethylamine, $CH_3CH_2\overset{..}{\underset{\displaystyle H}{\overset{\displaystyle |}{N}}}-H + H_2O \rightleftharpoons CH_3CH_2-\overset{\oplus}{\underset{\displaystyle H}{\overset{\displaystyle |}{N}}}-H + :\overset{..}{O}H^{\ominus}$

$$K_b = \frac{[CH_3CH_2\overset{+}{-}NH_3]\,[OH^-]}{[CH_3CH_2NH_2]}$$

9.14

I	0.10 M	0	0
C	$-x$	x^2	x^2
E	0.10 M^{-x}	x	x

$$K_b = 5.6 \times 10^{-4} = \frac{(x)(x)}{(0.10\ M^{-x})} \quad \text{Assume } x \text{ is small.}$$

$$5.6 \times 10^{-4} = \frac{x^2}{0.10\ M} \quad 5.6 \times 10^{-5} = x^2$$

$$7.48 \times 10^{-3}\ M = x = [OH^-]$$

9.15 $pOH + pH = pK_w = 14.00$

$pOH = -\log[OH^-] = -\log[7.48 \times 10^{-3}\ M] = 2.12$

$pH = 14.00 - pOH = 14.00 - 2.12 = 11.87$

9.16 analine

$$K_b = \frac{\left[\begin{array}{c}\overset{\oplus}{NH_3}\end{array}\right]\left[:\ddot{O}H^-\right]}{\left[\begin{array}{c}NH_2\end{array}\right]}$$

9.17

I	0.10 M	0	0
C	$-x$	x^2	x^2
E	0.10 M^{-x}	x	x

$$K_b = \frac{(x)(x)}{(0.10M^{-x})} \quad \text{Assume } x \text{ is small.}$$

$$3.9 \times 10^{-10} = \frac{x^2}{0.10M}$$

$$3.9 \times 10^{-11} = x^2$$

$$6.24 \times 10^{-6} M = x = [OH^-]$$

9.18 $pOH = -\log[OH^-] = -\log[6.24 \times 10^{-6} M]$
$pOH = 5.20$
$pH = 14.00 - pOH = 14.00 - 5.20 = 8.79$

9.19 % dissociation

Of the four bases in table 9-2, analine is the weakest base and ethylamine is the strongest.

$$\text{ethylamine } \% = \frac{7.48 \times 10^{-3} M}{0.10\ M} \times 100\% \approx 7.48\%$$

$$\text{aniline } \% \frac{6.24 \times 10^{-6} M}{0.10\ M} \times 100\% = 0.00624\%$$

9.20 1. primary 2. secondary 3. primary 4. tertiary 5. primary

9.21 1. 1-butylamine 2. N-methyl-2-butylamine 3. N-methyl-1-butylamine
4. N,N-dimethyl-2-butylamine 5. 2-butylamine

10 Organic Reactions

10-1 1.

2.

3.

10-2

$\text{C}_6\text{H}_5\text{–C(=O)–O–H} + \text{H}_2\text{O} \rightleftharpoons \text{C}_6\text{H}_5\text{–C(=O)–O}^{\ominus} + \text{H}_3\text{O}^{\oplus}$

10-3

$\text{H–C(=O)–O–H} + \text{H}_2\text{O} \rightleftharpoons \text{H–C(=O)–O}^{\ominus} + \text{H}_3\text{O}^{\oplus}$

acid   base   conjugate base   conjugate acid

10-4

$\text{H–CH}_2\text{–C(=O)–O–H} + \text{H}_2\text{O} \rightleftharpoons \text{H–CH}_2\text{–C(=O)–O}^{\ominus} + \text{H}_3\text{O}^{\oplus}$

acid   base   conjugate base   conjugate acid

10-5   When placed in water, weak acids react only from 1% to 10%. Most of the weak acid molecules remain intact as molecules. For example, if 100 molecules of ethanoic (acetic) acid were put in water, only about 5 molecules would actually react. This means that 95 molecules would remain intact and that (apart from water itself) acetic acid would be the predominant species.

10-6   acid-base ionic reaction

$\text{C}_6\text{H}_5\text{–C(=O)–O–H} + \text{NaOH} \rightarrow \text{C}_6\text{H}_5\text{–C(=O)–O}^{\ominus}\,\text{Na}^{\oplus} + \text{H}_2\text{O}$

Conj. base   Conj. acid

net ionic reaction

$\text{C}_6\text{H}_5\text{–C(=O)–O–H} + {:}\text{OH}^{\ominus} \rightarrow \text{C}_6\text{H}_5\text{–C(=O)–O}^{\ominus} + \text{H}_2\text{O}$

Conj. base   Conj. acid

10-7

$\text{C}_6\text{H}_5\text{–C(=O)–O–H} + {:}\ddot{\text{O}}\text{–H}^{\ominus} \rightarrow \text{C}_6\text{H}_5\text{–C(=O)–O}^{\ominus} + \text{H–O–H}$

10-8   Ionic: $\text{CH}_3\text{COOH} + \text{Na}^+ + \text{OH}^- \rightleftharpoons \text{CH}_3\text{COO}^- + \text{Na}^+ + \text{H}_2\text{O}$

Net: $\text{CH}_3\text{COOH} + \text{OH}^- \rightleftharpoons \text{CH}_3\text{COO}^- + \text{H}_2\text{O}$

10-9

$\text{CH}_3\text{CH}_2\text{C(=O)–O–H} + {:}\text{N(H)(H)–CH}_3 \rightleftharpoons \text{CH}_3\text{CH}_2\text{C(=O)–O}^{\ominus} + \text{H–}\overset{\oplus}{\text{N}}\text{(H)(H)–CH}_3$

10-10   1. $\text{CH}_3\text{CH}_2\text{CH}_2\text{C(=O)–O–H} + {:}\text{N(H)(H)–CH}_2\text{CH}_3 \rightleftharpoons \text{CH}_3\text{CH}_2\text{CH}_2\text{C(=O)–O}^{\ominus} + \text{H–}\overset{\oplus}{\text{N}}\text{(H)(H)–CH}_2\text{CH}_3$

2. 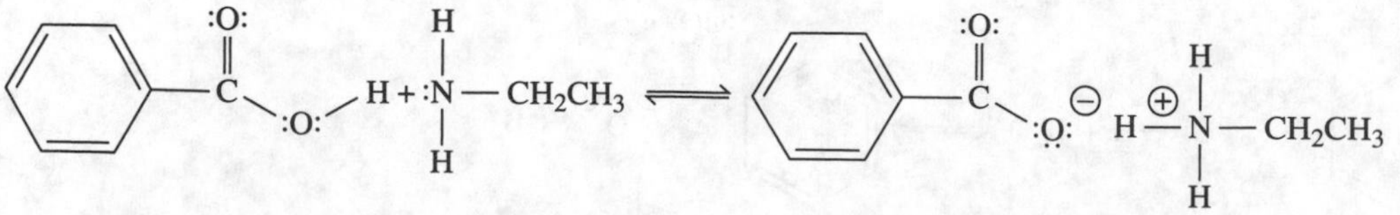 (reaction scheme with benzoate, ethylamine → benzoate anion + ethylammonium)

3. 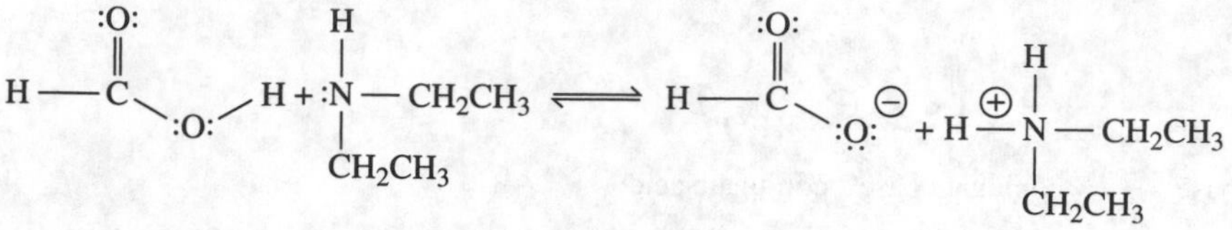 (reaction scheme with methanoate, diethylamine)

4. 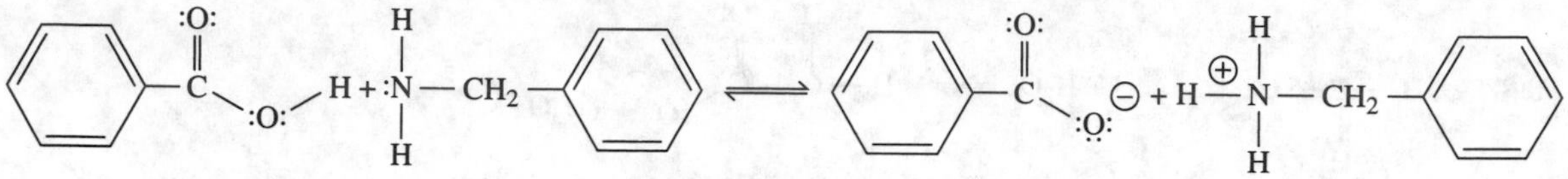 (reaction scheme with benzoate, benzylamine)

10-11

1. 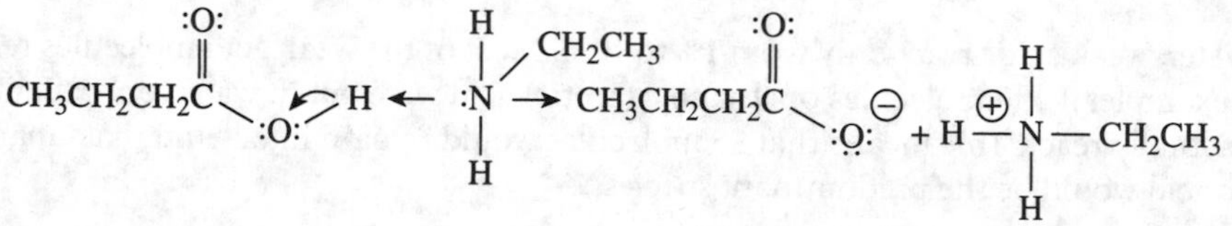 $CH_3CH_2CH_2C$(=O)(:O:H) ← :N(H)(CH_2CH_3)(H) → $CH_3CH_2CH_2C$(:O:)$^{\ominus}$ + H—$\overset{\oplus}{N}$(H)—CH_2CH_3 (H)

2. 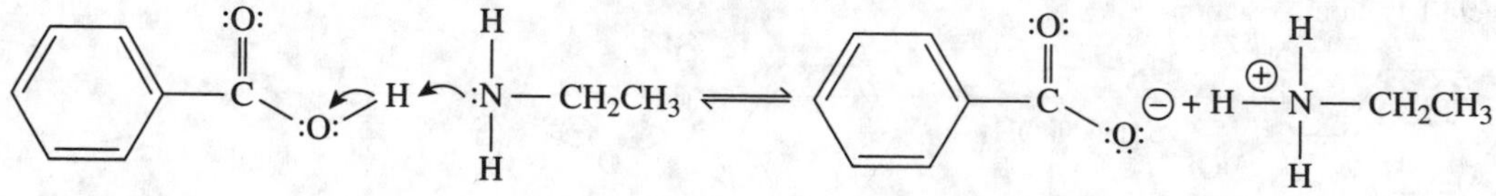 (benzoyl, ethylamine reaction) → benzoate$^{\ominus}$ + H—$\overset{\oplus}{N}$(H)—CH_2CH_3 (H)

3. H—C(=O)(:O:H) ← :N(H)(CH_2CH_3)(CH_2CH_3) → H—C(:O:)$^{\ominus}$ + H—$\overset{\oplus}{N}$(H)—CH_2CH_3 (CH_2CH_3)

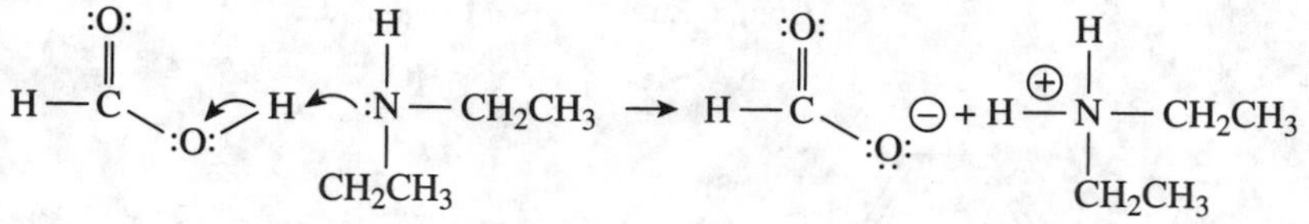

4. 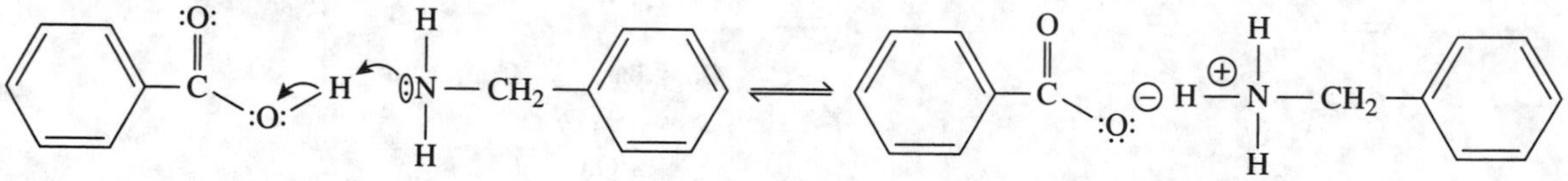 (benzoyl, benzylamine reaction) → benzoate$^{\ominus}$ H—$\overset{\oplus}{N}$(H)—CH_2—phenyl (H)

10-12

1. carboxyl, -COOH

2. 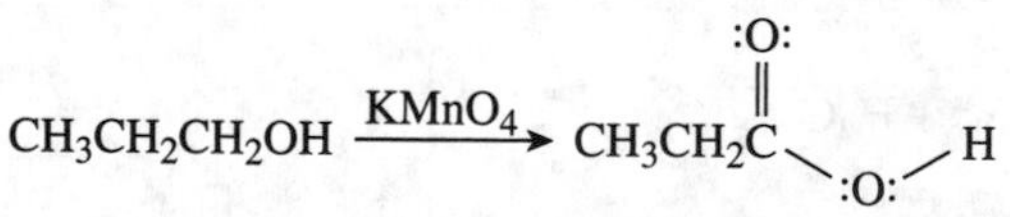 $CH_3CH_2CH_2OH$ $\xrightarrow{KMnO_4}$ CH_3CH_2C(=O)(:O:H)

3. 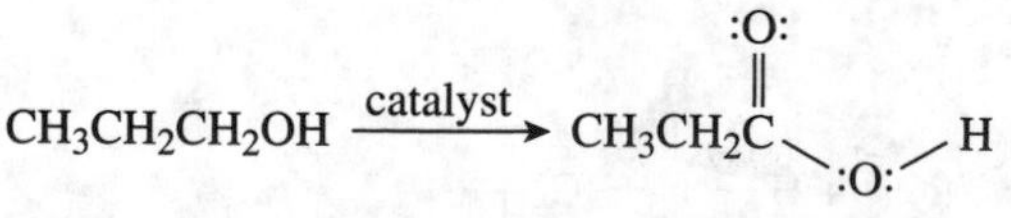 $CH_3CH_2CH_2OH$ $\xrightarrow{catalyst}$ CH_3CH_2C(=O)(:O:H)

4. methanoic acid

5. 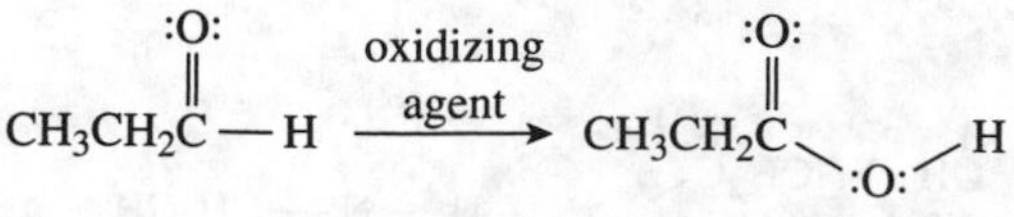 CH_3CH_2C(=O)—H $\xrightarrow[\text{agent}]{\text{oxidizing}}$ CH_3CH_2C(=O)(:O:H)

10-13 $CH_3CH_2\overset{\displaystyle :O:}{\overset{\displaystyle \|}{C}}\!-\!\overset{..}{\underset{..}{O}}:\!-\!H + CH_3OH \underset{}{\overset{H^+}{\rightleftharpoons}} CH_3CH_2\!-\!\overset{\displaystyle :O:}{\overset{\displaystyle \|}{C}}\!-\!\overset{..}{\underset{..}{O}}\!-\!CH_3 + H_2O$

10-14 1. $CH_3CH_2\overset{\displaystyle :O:}{\overset{\displaystyle \|}{C}}\!-\!\overset{..}{\underset{..}{O}}:\!-\!H + CH_3CH_2OH \underset{}{\overset{H^+}{\rightleftharpoons}} CH_3CH_2\overset{\displaystyle :O:}{\overset{\displaystyle \|}{C}}\!-\!\overset{..}{\underset{..}{O}}:\!-\!CH_2CH_3$

2. $H\!-\!\overset{\displaystyle :O:}{\overset{\displaystyle \|}{C}}\!-\!\overset{..}{\underset{..}{O}}:\!-\!H + CH_3CH_2CH_2OH \rightleftharpoons H\!-\!\overset{\displaystyle :O:}{\overset{\displaystyle \|}{C}}\!-\!\overset{..}{\underset{..}{O}}:\!-\!CH_2CH_2CH_3 + H_2O$

3. $CH_3CH_2CH_2\overset{\displaystyle :O:}{\overset{\displaystyle \|}{C}}\!-\!\overset{..}{\underset{..}{O}}:\!-\!H + CH_3OH \rightleftharpoons CH_3CH_2CH_2\overset{\displaystyle :O:}{\overset{\displaystyle \|}{C}}\!-\!\overset{..}{\underset{..}{O}}:\!-\!CH_3 + H_2O$

4. $CH_3\overset{\displaystyle :O:}{\overset{\displaystyle \|}{C}}\!-\!\overset{..}{\underset{..}{O}}:\!-\!H + CH_3CH_2CH_2OH \rightleftharpoons CH_3\!-\!\overset{\displaystyle :O:}{\overset{\displaystyle \|}{C}}\!-\!\overset{..}{\underset{..}{O}}:\!-\!CH_2CH_2CH_3 + H_2O$

10-15 1. $CH_3\!-\!\overset{\displaystyle O}{\overset{\displaystyle \|}{C}}\!-\!O\!-\!CH_2CH_2CH_2CH_3 + H_2O \underset{}{\overset{H^+}{\rightleftharpoons}} CH_3\overset{\displaystyle O}{\overset{\displaystyle \|}{C}}\!-\!OH + HOCH_2CH_2CH_3$

2. $CH_3CH_2CH_2\overset{\displaystyle O}{\overset{\displaystyle \|}{C}}\!-\!O\!-\!CH_2CH_3 + H_2O \underset{}{\overset{H^+}{\rightleftharpoons}} CH_3CH_2CH_2\overset{\displaystyle O}{\overset{\displaystyle \|}{C}}\!-\!OH + HOCH_2CH_3$

3. (structure: phenyl ring with $O\!-\!\overset{\displaystyle O}{\overset{\displaystyle \|}{C}}\!-\!CH_3$ and $\overset{\displaystyle C\!-\!OH}{\underset{\displaystyle \|}{O}}$ groups) $+ H_2O \overset{H^+}{\rightleftharpoons}$ (phenyl ring with OH and $\overset{\displaystyle C\!-\!OH}{\underset{\displaystyle \|}{O}}$) $+ HO\!-\!\overset{\displaystyle O}{\overset{\displaystyle \|}{C}}\!-\!CH_3$

10-16 1. $CH_3CH_2\overset{\displaystyle :O:}{\overset{\displaystyle \|}{C}}\!-\!\overset{..}{\underset{..}{O}}:\!-\!H + \underset{\displaystyle H}{\overset{\displaystyle H}{:N}}CH_2CH_3 \rightleftharpoons CH_3CH_2\overset{\displaystyle :O:}{\overset{\displaystyle \|}{C}}\!-\!\overset{..}{\underset{..}{O}}:^{\ominus} + H\!\overset{\oplus}{\underset{\displaystyle H}{\overset{\displaystyle H}{N}}}\!-\!CH_2CH_3$

2. $CH_3CH_2\overset{\displaystyle :O:}{\overset{\displaystyle \|}{C}}\!-\!\overset{..}{\underset{..}{O}}:\!-\!H + \underset{\displaystyle H}{\overset{\displaystyle H}{N}}\!-\!CH_3 \rightleftharpoons CH_3CH_2\overset{\displaystyle :O:}{\overset{\displaystyle \|}{C}}\!-\!\overset{..}{\underset{..}{O}}:^{\ominus} + H\!\overset{\oplus}{\underset{\displaystyle H}{\overset{\displaystyle H}{N}}}\!-\!CH_3$

3. $CH_3CH_2\overset{\displaystyle :O:}{\overset{\displaystyle \|}{C}}\!-\!\overset{..}{\underset{..}{O}}:\!-\!H + :\overset{..}{\underset{..}{O}}H^{\ominus}, k^+ \longrightarrow CH_3CH_2\overset{\displaystyle :O:}{\overset{\displaystyle \|}{C}}\!-\!\overset{..}{\underset{..}{O}}:^{\ominus} + H\!-\!\overset{..}{\underset{..}{O}}:\!-\!H + k^+$

4. $CH_3CH_2\overset{\overset{\displaystyle :O:}{\|}}{C}-\overset{..}{\underset{..}{O}}-H + CH_3OH \underset{}{\overset{H^+}{\rightleftharpoons}} CH_3CH_2\overset{\overset{\displaystyle :O:}{\|}}{C}-\overset{..}{\underset{..}{O}}-CH_3 + H_2O$

5. $CH_3\overset{\overset{\displaystyle :O:}{\|}}{C}-\overset{..}{\underset{..}{O}}-CH_2CH_2CH_2CH_3 \underset{}{\overset{H_3O^+}{\rightleftharpoons}} CH_3\overset{\overset{\displaystyle :O:}{\|}}{C}-\overset{..}{\underset{..}{O}}-H + HOCH_2CH_2CH_2CH_3$

10-17 2 > 1 > 3 > 4

10-18 1.

$H-C\equiv N: + $ (benzaldehyde) $\overset{NaOH}{\longrightarrow}$ (addition product) $\overset{H_2O}{\rightleftharpoons}$

2. $CH_3\overset{..}{N}H_2 + CH_3\overset{\overset{\displaystyle :O:}{\|}}{C}-H \overset{H_2O}{\rightleftharpoons} $ product

3. hydrazine $+ CH_3CH_2\overset{\overset{\displaystyle :O:}{\|}}{C}-H \overset{KOH}{\longrightarrow} CH_3CH_2-\overset{\|}{C}-H + H_2O$

4. $CH_3\overset{..}{O}H + CH_3CH_2\overset{\overset{\displaystyle :O:}{\|}}{C}-H \rightleftharpoons CH_3CH_2-\overset{\overset{\displaystyle :\overset{..}{O}H}{|}}{\underset{\underset{\displaystyle :\overset{..}{O}-CH_3}{|}}{C}}-H + H_2O$

5. (benzaldehyde) $\overset{NaBH_4}{\longrightarrow}$ (alkoxide) $\overset{H_3O^+}{\rightleftharpoons}$ (alcohol)

10-19 1. $CH_3CH_2COOH(aq) + KOH(aq) \rightleftharpoons CH_3CH_2COO^-(aq) + K^+(aq) + H_2O(l); CH_3CH_2COO^-(aq) + H_2O(l) \rightleftharpoons CH3CH_2COOH(aq) + OH^-(aq)$

2. $HOOCCH_2CH_2COOH(aq) + LiOH(aq) \rightleftharpoons HOOCCH_2CH_2COO^-(aq) + Li^+(aq) + H_2O(l); HOOCCH_2CH_2COO^-(aq) + H_2O(l) \rightleftharpoons HOOCCH_2CH_2COOH(aq) + OH^-(aq)$

3. $C_6H_5COOH(aq) + NaOH(aq) \rightleftharpoons C_6H_5COOH(aq) + Na^+(aq) + H_2O(l); C_6H_5COOH(aq) + H_2O(l) \rightleftharpoons C_6H_5COOH(aq) + OH^-(aq)$

4. $HOOCC_6H_4COOH(aq) + LiOH(aq) \rightleftharpoons HOOCC_6H_4COO^-(aq) + Li^+(aq) + H_2O(l); HOOCC_6H_4COOH(aq) + H_2O(l) \rightleftharpoons HOOCC_6H_4COO^-(aq) + OH^-(aq)$

10-20 When a weak acid reacts with a strong base, the conjugate ion that is formed reacts with water to re-form the acid and produces hydroxide, OH^-, which drives the pH to be basic.

10-21 Organic acids are weak because they do not ionize 100% the way the strong acids do. They remain primarily as molecules in water.